A Bridge Not Attacked

A Bridge Not Attacked

Chemical Warfare Civilian Research During World War II

Harold Johnston
University of California, Berkeley, USA

World Scientific

NEW JERSEY • LONDON • SINGAPORE • SHANGHAI • HONG KONG • TAIPEI • BANGALORE

Published by

World Scientific Publishing Co. Pte. Ltd.
5 Toh Tuck Link, Singapore 596224
USA office: Suite 202, 1060 Main Street, River Edge, NJ 07661
UK office: 57 Shelton Street, Covent Garden, London WC2H 9HE

British Library Cataloguing-in-Publication Data
A catalogue record for this book is available from the British Library.

A BRIDGE NOT ATTACKED
Chemical Warfare Civilian Research During World War II

ISBN 981-238-152-X
ISBN 981-238-153-8 (pbk)

Printed in Singapore by World Scientific Printers (S) Pte Ltd

CONTENTS

PREFACE

In this book, I tell novel true stories concerning highly talented civilian scientists in some unusual places during World War II, carrying out research on defense against poison gases. Most of these were graduate students, working under the direction of professors at the California Institute of Technology (Caltech) and the University of California (Berkeley). I focus on a small number of individuals, with in-depth study of these individuals and what they did. The general reader should be able to follow these mostly non-technical accounts of research on chemical warfare agents and to enjoy the many stories concerning, for examples, rural north Georgia during the Depression, rheumatic fever, Emory University in the late 1930s, the rising star of Caltech, the golden age of nuclear physics at Berkeley, and actions of these university scientists working together with army personnel on army bases. Scientists at Caltech and Berkeley did early laboratory work where the first job was to test how well our gas masks would protect soldiers against old and possibly new war gases. The gas masks miserably failed these tests, and the task was to develop better protection against toxic gases. High quality science at several universities went into this work, and the work succeeded in producing greatly improved gas masks. Later most activities were done outdoors to assess the effect of terrain and meteorological conditions on the travel and dissipation of toxic gas clouds. Professional chemists or meteorologists may be interested in some technical aspects of this book, but the technical material is almost entirely presented verbally and simply, and the general reader may, with no loss, skip the one chemical equation in Chapter 3, and the graphical figures and simple equations in Chapter 6.

As a first year graduate student in chemistry at Caltech, in April 1942, I joined Professor Roscoe G. Dickinson's war project in Division 10 of National Defense Research Committee (NDRC). In this project and a similar one at Berkeley, one young participant was a big healthy athletic extrovert, who was deeply trained in the physical sciences, and by age twenty-nine he was world famous in physics and in biology. Another was opposite in many ways: a skinny sickly loner, who was minimally schooled in science and mathematics. Of the ten principals on these two projects, one was killed by an accident while working with a poison gas in the laboratory. Another participant was proud of how he had defeated the draft system in an unusual way.

From Caltech we made micro-meteorological measurements in urban Los Angeles and on the beaches and deserts of southern California. From Berkeley they worked in the northern central valley of California, at Stinson Beach, and on Mount Shasta. An especially interesting story is how a horrible skunk stink saved the lives of the residents in a small village at the foot of Mount Shasta. In the fall of 1943, civilian scientists from Caltech, Berkeley, Northwestern University, and elsewhere joined a chemical warfare army group (Dugway Proving Ground Mobile Field Unit of the US Chemical Warfare Service) to carry out field tests with real chemical weapons in a semi-tropical forest in central Florida, including bombs dropped from airplanes. I worked at the Florida station until the end of the war. From February through September 1944, twenty-two men from several universities moved from Florida and elsewhere to a new army base in the jungles of Panama, where large scale poison-gas tests were conducted. Two of these young men volunteered to do especially dangerous work in New Guinea during the last year of the war.

The purpose of this book is to present an almost forgotten history of secret war research by chemists. I give mini-biographies of several of the participants from Caltech and Berkeley, Chapter 3 is a more extensive biography of one of them, and I present some autobiographical material. In support of this account, I have used and quoted at length material written in the 1940s: a humorous diary of a car trip from Pasadena, California, to Bushnell, Florida, letters written to their loved ones by four of these chemists, other letters including mine, technical reports, declassified reports, my class notebooks, my personal notebooks, papers, pictures, magazine articles, and newspaper clippings. I quote recently written contributions from five of the principal participants, and I interviewed others. I have pertinent technical books[#§], material from Caltech and Berkeley Archives, and material obtained via the Freedom of Information Act. I started writing this book in 1997.

On 27 June 1940, leading scientific statesmen and knowledgeable civilian political figures established the National Defense Research Committee (NDRC). A portion of its charge was that:

> The Committee shall correlate and support scientific research on the mechanisms and devices of warfare. . . It shall aid and supplement the experimental and research activities of the War and Navy Departments, and may conduct research for the creation and improvement of instrumentalities, methods, and materials of warfare.

Military officers participated in the identification of urgent problems. NDRC placed contracts with top university scientists to apply modern methods to a wide range of real problems.

This book does not pretend to give the full history of the chemical warfare work done by National Defense Research Committee (NDRC). That was done by W.A. Noyes, Jr., head of NDRC Division 10, who edited and contributed to a 1948 book,[#] which gives names, facts, and dates for NDRC projects in the chemistry divisions. In Division 10, there were seventy-seven projects placed in eighteen research universities and five industrial laboratories. There were twenty divisions in NDRC, each with its own many projects. Most of these projects probably involved scientists and people as interesting and talented as those in my sample. From lists of names in this book, I estimate that about 700 civilian chemists were engaged in NDRC war research. Also in 1948, the Chemical Warfare Corps published a book reviewing its activities throughout the war,[§] where the NDRC contributions are strongly acknowledged.

NDRC mobilized research universities of the country to contribute to the war. I give here a reminder of this part of history with an intimate example.

[#]*Chemistry. A History of the Chemical Components of the National Defense Research Committee.* Edited by W.A. Noyes, Jr. An Atlantic Monthly Press Book, Little, Brown and Company, Boston, 1948, 524 pages. NDRC was one branch of the Office of Scientific Research and Development (OSRD). In 1997, Professor Charles Parmenter gave me his copy of this book, for which I am deeply grateful.

PUBLISHER'S NOTE: "Under the terms of the contract for the publication of *Chemistry* . . . the publisher has agreed to waive its right under the copyright . . . after ten years from the date of publication of such volume. Thereafter the volume in question will be in the public domain and dedicated to the public."

[§]The Chemical Warfare Service in World War II, *A Report of Accomplishments,* REINHOLD PUBLISHING CORPORATION, 330 West Forty-second St., New York 18, U.S.A., THE CHEMICAL CORPS ASSOCIATION, Washington 1, D. C. (*Manuscript was released to the public on 20 February 1948*).

ACKNOWLEDGMENTS

I am deeply grateful to the many others who wrote parts of this book or contributed documents or photographs. In a special way I am grateful to Calvin Kytle, my good friend since he and I were freshmen at Emory University in 1937–1938, who read the entire manuscript and made wise suggestions about how to reduce, revise, and focus the book. I reworked most of the material in light of his suggestions. I also thank Karen Krushwitz, David Chewning, Kate Gilpin, and Professor Bill Gruber for their constructive advice.

Helena Ruben saved Sam Ruben's letters and gave them to her son George in 1997. George, with Helena's concurrence, gave me permission to use whatever portions of the letters I chose. George Ruben also provided the photographs of Chapter 3, and he gave me legal documents, newspaper clippings, magazine articles, and other written material. George's sister-in-law, Ada Ruben, contributed genealogical information about Sam's ancestors.

John Otvos' family saved his letters from Panama, and he permitted me to choose portions from them. Also, John had prepared for his children an account of his early life in Hungary, his family's moving to this country, his education, and in reduced detail his experiences after he obtained his Ph.D. degree from Caltech. He let me select large portions of this material for Chapter 8. John permitted me to use pictures of "Egbert" or the Dickinson meter from his Ph.D. thesis, and he gave verbal accounts of his work in Panama.

During 1944, when NDRC graduate students were working with the army in Panama, Bob Mills and Rene Scott exchanged letters, and Rene saved all of them. I quote many of Bob's letters and some of Rene's letters in Chapter 5. Bob's letters give no information about their secret war work, but they give an excellent picture of their living, working, and recreational activities in the jungles of Panama. Rene's letters are rich with gossip concerning Caltech: lovers, weddings, parties, the coming and going of people, and baseball. Rene also gave me Bob's diary of the trip from Pasadena to Bushnell, Florida, and her write-up of one inning of a baseball game.

In two long letters, John Thomas charmingly tells about the NDRC group at Berkeley during 1943–1944, especially concerning Sam Ruben's project. He prepared a fascinating list of topics that he would write up concerning the war work in Panama, but he was unable to do this because of failing health and his death in 2001.

Andy Benson contributed a large amount of insight to Chapter 3 and some written material. Andy was not employed in the NDRC projects, but he worked closely for two years with Sam Ruben on pioneering photosynthesis work and biochemical experiments with rats.

Bill Shand was a brilliant scientist and a skilled and daring mountain climber. He took many pictures of the San Jose Island expedition. He finished his Ph.D. degree at Caltech after the end of the war, was appointed Instructor at Berkeley, and died in a car accident as he sped through Nevada on his way to climb in the Rocky Mountains. Bill's parents gave most of Bill's Panama photographs to Bob Mills. When Bob died, the pictures were the property of Rene Scott Mills, his wife. Rene (rhymes with clean) gave me the right to include any of these pictures in this book, and they contribute heavily to Chapter 5.

Professor David Templeton wrote a letter describing Helena Ruben's work in his laboratory.

I received a large amount of background material through the Freedom of Information Act, from the Caltech Archives, and from the Berkeley Archives. I copied and included some photographs, letters, reports, and documents from these three sources.

I am grateful to DeLorme, Street Atlas USA, for permission to print their maps as Figure 1 in Chapter 4, and to the Archives of CBS 60 Minutes for Figures 8 and 9 of Chapter 5.

I am especially grateful to Professor Charles Parmenter for giving me his copy of the book by W. A. Noyes, Jr., 1948.

CHAPTER 1

HAL

ANCESTORS

Doctor Medicine Johnston, Jr., at age nine, saw and heard the Battle of Kennesaw Mountain, Georgia, as the Union troops forced their way toward Atlanta. His father was a tenant farmer living in southwest Cherokee County, Georgia. At age fourteen, D.M., Jr., had to take care of his mother and sister when Doctor Medicine Johnston, Sr. died in 1869. The name, Doctor Medicine, was on the birth certificates, and it poses some mystery. In later years, D.M., Jr. had his legal name changed to J H Johnston, no period after the initials, no meaning to the letters. He worked as a farm laborer and in the gold mines of north Georgia, borrowed a hundred dollars to buy a small farm near Moon's Crossing, in time repaid the debt, and slowly increased his farm to three hundred acres. In 1888, he sold the farm and set up J H Johnston Co., a general store in Woodstock, Georgia, thirty miles (50 km) north of Atlanta. In addition to the store, over the years he prospered, became a cotton buyer and seller, founder of a small bank, land investor, and local patriarch. He built a large house on a narrow lot in town; the house had six tall, white Ionic columns on the front porch and heavily used brass spittoons in the living room.

As I knew him, Grandpa Johnston was tall, old, and severe. He had white hair and a big white mustache. He boasted that he had gone to school for only six weeks in his life, and he regarded college as a waste of time and spoiler of the young.

When the men of the family came back from hunting or fishing, they dumped their dead ducks or fish or whatever into the kitchen sink, and "poor little Grandma" had to pull out the feathers or scrape off the scales and clean them, as well as cook them. In later years, she had help from a house servant. Several times a year on holidays or on a Sunday, Grandfather and Grandmother fed their grown children and their grandchildren a huge and varied midday dinner, typically including two or three meats, such as,

---

Some of this section is based on: Harold Johnston, Autobiographical Sketch, *Journal of Physical Chemistry* **A105**, 1388–1390 (2001).

turkey, ham, roast beef, occasionally venison or opossum, on one occasion bear, and many vegetables. The men ate first, and the grandchildren were warned most strongly to behave even though they were hungry. The men sat around talking politics long after eating their fill, and only when they had finally gotten up did the women and children get their turn. Grandma warmed up some of the food, but much of it was cold, and she sat around eating nothing herself until the women and children had almost finished.

* * *

The Cherokee Indians had their own written language and a newspaper. They both hunted and farmed, were rapidly adapting to the ways of the whites, and wanted peace and co-existence with them. In the early 1830s the Cherokee Indians were cruelly driven out of Georgia. The land was split into forty to two-hundred acre lots and given by lottery to settlers. My maternal great-great-grandfather, Joseph Stallsworth Dial (1793–1867), was born in South Carolina, moved into Cherokee County a few years after the lottery was over, bought many of these lots as some settlers gave up and left, and accumulated extensive acreage east of Woodstock. My maternal great-grandfather William Choice Dial, Sr. (1826–1902) inherited a quarter of this land, and he comfortably supported his large family by continuing to grow cotton. Facing opposition from the parents of the girl he wanted to marry, my grandfather William Choice Dial Jr. (1867–1940) at age twenty-one eloped with Dollie Gresham, age eighteen. Thirty months later Dollie died of pneumonia, leaving two baby daughters, fifteen months old Florine (later my mother) and little Dollie, age six weeks. My mother was reared by her aunt, Elizabeth Dial Latham, who ran a boarding house in Woodstock. In 1902 Grandpa Dial inherited and farmed one-fourth (160 acres) of his father's estate and remarried. Grandpa Dial lost his land to the bank, and his home burned down. My father bought a small house in Woodstock and gave it to him; he then lived in Woodstock and supported himself by doing odd jobs. Over the generations, some go up and some go down.

HOME, PARENTS AND BROTHERS

A photograph from our family album, taken sometime between 1915 and 1920, shows downtown Woodstock, (Figure 1.1). The road was unpaved. The people were owners or workers in the stores. J H Johnston general store was at the center of the picture and to the left of Dean's Drugstore.

Figure 1.1. Downtown Woodstock, Georgia, about 1915. From family picture album, unknown photographer.

The economy of downtown Woodstock was based on cotton farming in the surrounding area. The depression, invasion of boll weevils, and widespread depletion of top soil impoverished the whole region.

Here, I quote myself from an "Autobiographical Sketch" (Reprinted in part with permission from *Journal of Physical Chemistry* **A105**, 1388 (2001). Copyright 2001, American Chemical Society):

> Thirty miles NNW of central Atlanta, Woodstock had a population of 420 in 1930, 419 in 1940. Its economy was based on cotton farming. In 1930 there were more horses, buggies, and wagons than cars. Across Main Street from the railroad line, our family home, on a one acre lot, included a garden, chickens, a cow, two pigs, and flies. I remember when the first paved road and first electric power for homes came to Woodstock in the 1920s.
>
> My father Smith Lemon Johnston wanted to go to college and become a Methodist Minister, but Grandfather Johnston forced him not to go to college but to work in the family business. When he reached 21, my father paid his own way at Young Harris College for one year, but then he surrendered and became a partner in the J H Johnston Co. and, later, in the Bank of Woodstock and in the Cherokee Rope Mill. He was an extremely active Lay Leader of the Methodist Church and went to

conferences one or more times every year. He laid down the law in the family. For serious offenses, he administered the lashings after he got home from work. My mother was an artist and scholar by nature, but was strongly limited in finding means to express it. She was a talented pianist, loved to read good books, and was serious about her ceramic art creations. A relative had promised to pay her way through a four-year college, but when she was ready for college, he had died without a will. Her Aunt and Uncle paid her way for two years to Georgia State College for Women in Milledgeville, and she taught school for a few years before marrying my father. She introduced her children to books by reading to them after they were six months old, and encouraged them to read for themselves later.

I have three brothers, each distinguished in his own field: Smith L. Johnston, Jr. (1918–2000) graduated from Emory University in 1939, worked in an Atlanta bank (1939–1942), Private to Captain in the U.S. Army (1942 to 1945), and over the years became a partner and later president of the Bank of Woodstock, principal owner of J H Johnston Company, real estate investor, and local benefactor. Richard Johnston (b. 1923) received the B.S. Degree in Textile Engineering from Georgia Tech, worked for a time in a cotton textile mill, served in the U.S. Air Force (1943–45), was Senior Research Scientist at Georgia Tech until retirement; and now he lives in Atlanta. William Johnston (b. 1932) received the Doctors Degree in Veterinary Medicine from the University of Georgia, practiced veterinary medicine until retirement, and now lives in Woodstock.

Each of my parents had strongly wanted to go to college, and each had been disappointed. In spite of hard times during the depression, they supported all four of their children through college.

CHILDHOOD

At age eight, I had one sin preying on my mind. My brother, Smith, Jr., three years older than I, played baseball with the boys across the street, Ralph and Lewis, who were respectively one and three years older than Smith. These big boys collected cigarette butts off the street, recovered the residual tobacco, rolled it in brown wrapping paper, and smoked it, literally behind the barn. The family home site was about half a hectare lot, with the house at the front, the barn, pig pen and chicken pen at the rear, and the garden in between. One day I accidentally discovered the boys in the forbidden act of smoking. The big boys then suddenly became friendly, complimented me, invited me to sit down with them and to take a little smoke myself, which I did. So, having proudly joined the club I could no longer tell on them. On my own, I tried other things. Ground coffee rolled in brown wrapping paper made a good smoke. Dried corn silk was terrible, it burned too fast, scorched my mouth, and left me gasping. The stems of grape vines between the joints smoked pretty well. Later, I introduced smoking to my brother Dick, three years younger than I. He has a great sense of humor and a streak of daring, but we mostly restricted our smoking to grape vine stems and rabbit tobacco. A few months after these experiments with smoking, I felt a sudden shortness of breath while racing with other boys. Some teachers and the preacher had said that if a boy smoked, it would stunt his growth and give him shortness of breath. I was convinced that my smoking caused it. As time passed, the shortness of breath got worse, but I could not tell anyone about it, because it was my own fault, it was my own sin. At home I was regarded as lazy, and with peers I was inept in sports.

The Woodstock Public School sat on top of a large hill, which sloped gently down to the highway. Three broad flat terraces had been cut into this hill; the one next to the highway had basketball hoops but could be used for other sports, the largest terrace was devoted to baseball, and the one on top had a few shade trees and wooden benches. On three sides of the school, the heavily wooded hill sloped steeply downwards, and at the bottom to the north there was a large level field for full scale baseball or for the circus which visited in the summer. A dirt road from and to the highway encircled the school and the terraces. The superintendent, Mr. Eugene Booth, lived next door to the school, and the big schoolboys brought water from his well, providing one bucket of water and one dipper to each of the five school rooms. A coal burning stove in each room provided heat in the winter, and a can of water on the top of the stove slowly boiled to build up the humidity. There were wooden four-holer out-houses part way down the steep hill, one for the boys and one for the girls.

The two-story building had grades, one through ten, but it had only five classrooms. Each pair of classes shared one room and one teacher. Mr. Booth taught grades nine and ten, was an excellent teacher, tutor, and adviser, tried to inspire the students to study hard and prepare themselves for further education.

Each day there was a twenty minute recess and an hour off for dinner. When it was time to play ball at Woodstock School, the older boys gathered together. One star ball player would pick up a baseball bat and call out, "Play taw." Another boy would step up and say, "Play taw." The first held a baseball bat upside down and tossed it to the other, who caught it above the middle point. The two then took turns in grasping the bat, hand over hand, until the end of the bat was reached. Whoever covered the end of the bat got to make the first choice from among the players assembled. Then each took turns selecting the next player, and each tried to get the best players on his team. Everybody had to be picked eventually; this was the school rule, and usually I was picked last and teased about it. On one occasion, two girls decided that they wanted to play baseball with the boys, the boys told them to go away, but they wouldn't go away. Finally, the boys let them play, but they picked the girls last. The girls were actually pretty good players. A few days later the girls were picked next to last, and I was picked last. That really became a joke being selected after the girls. Many took up the joke. They teased, and some called me a sissy. At about age twelve, I quit trying and turned to walking in the woods.

I loved to explore the extensive forests beyond the cotton fields. Noonday Creek was about half a mile (0.8 km) from my home and Little River was less than two miles (3 km) away. The trees were mixed pine and hardwood. By finding and reading pamphlets, and by asking people questions, I learned to identify dozens of different trees and shrubs. It was fun to climb trees. From my perch high in a tall tree, I picked, then pressed to my lips the tough-skinned muscadine wild grape and squeezed it with my fingers to pop the sweet stuff into my mouth. With my teeth I separated out the seeds to spit them out. The falling seeds sounded like rain against the leaves of the undergrowth. My younger brother Dick often joined me in exploring the woods. Dick could climb trees almost as well as I could.

In the cotton field just beyond Noonday Creek, there was an area where I, alone or with my brother Dick, found Indian arrowheads, other obsidian tools, pestles and mortars, and broken pieces of pottery. After a farmer had plowed the field in the spring, rains washed the top soil and exposed these relics. I collected and classified these findings into cardboard boxes that eventually filled up a large wooden box.

* * *

We went as a family to the Methodist Sunday School and to church services afterward. No one could read the Sunday funny paper until after church. We said "grace" before each meal. In the summer, the church had a week or two of "revival meetings" during which there were sermons twice-a-day. These were strongly emotional events. The regular and visiting ministers raised their voices, told harrowing stories, and dwelt on sin, hell-fire, and eternal damnation. After preaching in this vein for almost an hour, the visiting preacher stepped down from the pulpit to the altar, held out his arms and continued to intone in the same vein. After a week of this, I, who was then just ten years old, was overwhelmed. I joined the line that was going down the central aisle. When I got to the preacher, I blubbered something, and the preacher loudly proclaimed, “God bless you, my boy. Take your place at the altar.”

SICKNESS

When I was thirteen, I was sick in bed with fever, swollen knees, and other things. After the doctor left, I could hear my mother weeping as she walked around downstairs doing her chores. From the questions the doctor asked of me that first day, I thought it might be tuberculosis. Many people around Woodstock, young and old had died of TB, but I was sure that going to a sanitarium for a year would certainly cure it. I pictured myself returning from the sanitarium after a year, suntanned and strong, surpassing my classmates by a star performance in basketball. I asked my mother if it was TB. She said it was nothing that bad, that I simply had a slight infection in the heart called rheumatic fever, and could go back to school after a few weeks. She said that I would have to take it easy for a year or two, and implied complete recovery after that. A year or two seemed like an awfully long time, but I assimilated and accepted my mother’s prognosis. In a mirror, I could see a bulge come and go under the left nipple as my heart beat, and it didn’t do that before. During the next summer we went to the Atlanta doctors. They found I had an enlarged heart, fast irregular pulse, and a heavily leaking mitral valve. They advised me not to race after a streetcar if it was pulling out before I got there but to wait for the next one. They insisted that I not engage in any competitive sport, and avoid climbing stairs.

Toward the end of summer of 1934, I was in the bathroom, sitting on the toilet, and idly reading a thin local magazine the family subscribed to. On the back page, there was an article about rheumatic fever, which I read. The article stated that there were now many fewer cases of rheumatic fever than during the war, but that it was still a significant disease in Georgia. Mostly it hit children between the ages of five and fifteen, and the article briefly recounted symptoms and the various outcomes. After reading, in the last paragraph, that the average survival time of those who had rheumatic heart-valve disease was fifteen years after the first attack, I became bodily stricken with fear of death, but in about thirty seconds I had gotten over the fear and begun to make plans as to what to do. They had not told me; I resolved not to tell them that I knew of my prognosis. I perceived that I would have more freedom this way and a better chance to do what I wanted.

If my shortness of breath since age eight was due to undiagnosed rheumatic fever, then by adding eight to fifteen I obtained twenty-three years as the probable life time for me; but if I added fifteen to thirteen I could expect to live until about twenty-eight. One might wonder how my doctor could have not noticed heart damage from rheumatic fever for five years. We didn't see a doctor often, and he didn't have a stethoscope. He listened by putting his ear on my chest, and I suppose he didn't do this except when he thought he had to. Anyhow, I decided to take age thirteen as the starting point of the calculation and rounded out to give myself thirty years.

* * *

Brother Smith swapped an old bicycle for an old twenty-two gauge single-shot rifle. Smith let me buy six cartridges from him and shoot the real rifle in target practice against a steep red-clay bank. Later, when Smith and our mother were away, Dick and I found the hiding place of the rifle on a top shelf of a closet. We took the rifle into the backporch bedroom. We cocked the trigger, aimed at imaginary monsters in the apple tree, and snapped the trigger of the empty rifle. I still had three cartridges and showed Dick how to load and unload the bullet. I took the rifle and proposed to play a game, "I'll shoot you with the empty rifle, and you fall dead. Then, I'll let you do it to me." I cocked the trigger, aimed the rifle at close range to Dick's forehead, and prepared to pull the trigger. But then I paused, pointed it away, relaxed the trigger, and breached the rifle. A live bullet was ejected! I thought to myself with profound distress, "I nearly killed my brother." I quickly returned the rifle to its hiding place, rushed out of the house, dug a hole in one inaccessible corner of the garden, and buried the three bullets in the red soil. Then I ran to the live oak tree at the far side of the garden; the deep shade excluded weeds, and the limbs of the oak came down to the ground, creating a secret hiding place. I sat on the leafy ground and silently repeated over and over: "Thank you, God. Thank you, Jesus. You saved me from killing my little brother." The feeling of sin would not go away. Repeatedly I visualized what would have happened if I had pulled that trigger. I vowed to read the Bible every day and to read it all. I decided to become a Methodist preacher. I never told anyone about the horrible crime I almost committed, nor about my vow to read the Bible, nor about the decision to be a preacher. On August 23, 1936, I completed reading the entire Bible, begats and all; on February 10, 1937, I completed the second reading of the Bible.

HIGH SCHOOL

I rode the bus for twelve miles (19 km) from Woodstock to Canton for two years of high school, grades ten and eleven. Students assembled in their home room, at the first bell the home-room teacher took the roll, and at the second bell the students marched in a line to the school auditorium, where Mr. Cash, the school superintendent, presided. Mr. Cash was old, and his skin was gray. He called for and led the morning song, made his announcements, and closed with the morning prayer.

The students went separate ways, each to his or her own classroom, which was typically other than the home room, and at the next bell moved from one classroom to another. During both years, the home room of my class was on the second floor, but to avoid my having to climb stairs I was assigned as home room a seat in one corner of the big library room. As my class walked down the stairs into the auditorium each morning, I slipped into the line and proceeded with them. I fulfilled my high school requirements from courses offered on the first floor. The librarian was a beautiful girl in her early twenties, it turned out that she too had rheumatic fever heart damage. The two of us discussed many things together, including the new book, *Gone with the Wind*, and we became good friends.

* * *

Most students brought lunch to school, ate rapidly, and played baseball or did other activities during the remaining lunch time. To build up weight as the doctor recommended, I walked over to Hotel Canton and had a hot dinner every day, which cost thirty-five cents and took up the hour. I was recognized as "the sick kid," no sports participation was expected, most boys politely ignored me, but several girls were friendly.

MY EARLY EXPERIENCES WITH CHEMISTRY

One Christmas I had received a chemistry set and enjoyed carrying out the examples given in the instruction sheet and in making other experiments. My favorite reading at that time was *Boys Life* magazine, and one story that especially impressed me is summarized here as I remember it:

> A well-to-do old man in the presence of three companions stopped in front of an old unoccupied house he owned. He fiercely reprimanded the young man there whom he had hired to clean up the yard and tested the yard water hydrant connected to a hose, "to be sure it was turned off." Shortly after the four men left, the yard boy saw purple flames in the front room of the house, reported the fire, and, on the testimony of four witnesses, was arrested for arson. His motive was said to be vengeance on the man who had reprimanded him. The old man collected fire insurance on his run down, burned down house. The yard boy insisted he was innocent. The local pharmacist, who in college had majored in chemistry, tended to believe the boy, marveled at his account of the purple flame, traced the burned off hose up to the ashes of the house, and found the metal end piece inside. The chemist collected ashes at the site of the metal end piece of the hose, analyzed the ashes, and found a large amount of potassium hydroxide near where the fire started. Ashes elsewhere showed only traces of potassium hydroxide. Water sets metallic potassium on fire and forms potassium hydroxide and flaming hydrogen gas. The police detective, alerted by the pharmacist, found that the old man had bought sixteen ounces of metallic potassium. The three witnesses admitted that the man had tested the water hydrant. At the trial, the yard boy was found not guilty, and the old man was then arrested for arson and fraud.

I was fascinated by the idea that water could set something on fire. I consulted the *Encyclopedia Britannica* about potassium and its chemical little brother sodium, which also burns in water. From a small advertisement in *Boys Life*, I ordered three ounces of metallic sodium for $1.65, which arrived immersed in kerosene in a metal container. Sodium catches fire on contact with water, and three ounces tossed into a lake would have made a good brief Fourth-of-July display, but I made better use of it than that. After some experimentation, I cut off a piece about the size of a small lima bean, dropped it in a small flat can with a little water on the bottom, covered the can by tipping over it a large inverted metal wash tub, and quickly ran away. The sodium reacted with water to give off hydrogen gas and sparks. The hydrogen rose toward the top of the tub and built up until it reached the critical explosion limit, and then a sodium spark ignited the air-hydrogen mixture, sending the tub up as high as two-story house with a loud bang and a column of irritating white smoke of sodium hydroxide. In this way the sodium lasted for almost two years and provided much fun.

* * *

Believing that I would not live beyond the age of thirty, I wanted to accomplish something in life and decided to write a Great Novel. I took a journalism course in high school and was editor of the student newspaper in my senior year. I received a $200 scholarship to Emory University for editing the best student newspaper in a high school of its size. The scholarship covered almost a third of total college expenses for one year, which were $676.50 (I still have the page from a ledger that totaled up my college expenses). From larger high schools two other students received journalism scholarships, and the three editors, David Chewning, Calvin Kytle, and I, became long-term good friends.

EMORY UNVERSITY

I entered Emory University in the fall of 1937 to major in English Literature. Dean Rece lectured to the newly arrived freshman class, laying down the rules of Emory and of its honor system, telling us we were to sit in our assigned numbered seat in the Glenn Memorial Chapel every Friday with attendance kept by observers in the balcony, and that we had to understand and follow the important announcements that Emory officers would make.

The members of the freshman class were given appointments for a medical examination. According to verbal instructions, all of us in our group got in line, nude, and filed past three doctors. Each student handed in a paper on which he had filled out a medical questionnaire. The doctors asked questions and made some simple examinations. At the end of the examination I was told that my life expectation was between thirty and forty years. I regarded the "or forty" as good news, but from what I had read, I really did not believe it.

I had taken chemistry and physics in high school, the chemistry course had been interesting, but the physics teacher didn't understand the material, and spent class time dully reading from the textbook. I was required to take one science course in college, and my playing with sodium and the interesting high school course probably led to my choosing chemistry instead of biology or physics. The freshman chemistry course was taught by the head of the department, Professor J. Sam Guy, who was an excellent and enthusiastic teacher.

During my first year at Emory, I went to a different place of worship almost every Sunday and read about different Protestant Sects, the Roman Catholic Church, the Muslim religion, Buddhism, and the religions of primitive people, including the one-volume version of Frazer's *The Golden Bough*. While reading one afternoon late in my freshman year at Emory, I dropped my secret plan of becoming a Methodist Minister.

As some saw it, I got pretty wild during the spring and summer of my freshman year. I quit going to church, quit reading the Bible, refused to follow the doctor's instructions, and made a game of climbing multiple flights of stairs. I snickered upon reading Aristophanes and Boccaccio. I drank some wine. During the summer vacation in Woodstock, I had trouble finding things to do and complained about it. My great-aunt Lizzie, effectively my grandmother, visited at length that summer, and I decided to play a trick on her. Any time I sat down to read, she could see that I was doing nothing and would come up to start a conversation. One time although I had just finished reading a book, I sat in a comfortable chair and started reading it about one-third of the way through. To be sure, Aunt Lizzie soon appeared, "Harold, dear, what are you reading in that great big book?"* I held the book up so she could read the title: *The Origin of Species and the Descent of Man*. Aunt Lizzie then exclaimed with horror, "Darwin, Darwin, oh, oh, oh, oh." and she quickly left and told my father what wicked things I was reading; but my Father's only comment to me was that he was more interested in where we were going than in where we came from.

* * *

In high school, I played a clarinet. I learned about the frequencies of the musical notes and on semi-logarthmic graph paper plotted the frequencies against time for songs in the church hymn book. I wanted to compose music by drawing graphs, but this project was unproductive. While in bed in the winter of 1934, I was given sole use of the family radio. One Saturday afternoon, I called my mother in and exclaimed, "They are having a play on the radio, but instead of talking, they are singing. Have you ever heard of anything like that?"

"Oh yes," she said, "this is called opera, Grand Opera."

The speaker described the story in some detail, it was a new opera sung in English, but I could understand only a few of the words. Operas on

---

*I distinctly remember the event and the nature of the discussion with Aunt Lizzie, but of course, I do not remember the exact words that were used. Consistent with what I recall, I construct a conversation and put it in quotation marks. I follow this practice throughout the book.

subsequent Saturdays usually were not in English, but it was the music that was appealing. I became a devoted amateur consumer of music. Music speaks nonverbal things directly to the brain, and the brain responds, "This Is Significant." After I got out of bed, the radio was also used by the rest of the family. At Emory University there were non-credit music appreciation seminars; the first time I heard the records of Beethoven's Fifth symphony, I was overwhelmed. A friend and fraternity brother, Ernest Abernathy, had his own record player and some records, mostly Beethoven's symphonies, which I liked almost as much as Ernest did. I bought a record player and started to collect records, but they were expensive. I bought some single records, each having excerpts from a single opera, and supplemented them with the libretto of the opera, both in English and in the original language. Records usually blended the many instruments of a symphony orchestra into a single voice, occasionally more, but I found that Beethoven's quartets were almost always resolved into two or three voices, which added interest. I bought the scores of all sixteen Beethoven quartets, and records of as many Beethoven quartets as I could afford; and I followed the scores while playing the records. As an Emory student, I could occasionally serve as usher to touring symphony orchestras or singers, and after the ticket holders were placed, the ushers could take free any empty seat in the house.

* * *

My older brother Smith worked every summer between college years in order to make some money to help pay for his education the next year. In view of the problems of my idle summer after the freshman year, my parents found a job for me the next summer. In the rural areas of North Georgia of that time, there were two especially busy working periods. In spring and early summer the ground had to be plowed, the cotton planted, and the weeds hoed until the cotton plants were strongly established. In the fall, the cotton had to be hand picked and hauled to the cotton gin. The school system in the farming areas adapted to this schedule. The children went to school for two months in the summer and for six months in the late fall, winter, and early spring. I was offered a teaching job for the summer at thirty dollars per month for two months, provided I received a teaching certificate, which could be obtained by taking three courses in Education at Emory. At Emory, every course was for five units and had five lectures each week. The normal load was three courses per quarter, a four-course load was allowed for anyone with a B average or better, and a five-course load was not allowed. After my second quarter at Emory, I took four courses every quarter; and since I had already picked out twelve neat courses I wanted to take in the sophomore year, I petitioned for permission to take five courses, but the petition was denied. In my three courses in Education, I found nothing that would help me be a better teacher: General Education was about the history of administration in primary schools; there was some dull Economics. Educational Sociology was about sociology from the point of view of the professional educator. However, I got through the courses and received the Teaching Certificate.

OAK GROVE SCHOOL

At Oak Grove School, I had my first teaching experience. The school was four miles (6.4 km) of unpaved roads from Woodstock and close to where J H Johnston had his three hundred acre (120 hectare) farm. Usually, my mother drove me to and from the school, or I rode one of Grandpa Dial's horses, but for several days I had boarded at a farmhouse near the school, because Noonday Creek had flooded the road; Kevin Boston and I had tried to wade through flood waters to reach my mother's car on the other side, but the water was too deep (Figure 1.2(a)).

Figure 1.2a. Hal J. and Mr. Boston unsuccessfully trying to cross flooded Noonday Creek. Photograph by Mrs. Florine Dial Johnston in family album taken in 1939.

The students walked to school, some of them as much as two miles (3.2 km) each way. There were three teachers, Mr. and Mrs. Kevin Boston, and "Mr." Johnston. There were three classrooms, grades one through three, four through six, and seven through nine. Mrs. Boston taught all classes for the first three grades, and classes four through nine were divided between Mr. Boston and me. I taught English, spelling, mathematics, civics, and, to three students in the ninth grade, Latin. Mr. Boston said he had to teach the science and history courses and warned me to be careful; the previous year a student's irate father had knocked down the science teacher and "stomped him," because of some sacrilegious thing the teacher had said in the science class. Some of the ninth-grade students were bigger than I and almost my age, which was eighteen, another reason to be careful. That year, for the first time, the State of Georgia supplied free textbooks to students, but the books were to be returned at the end of the year, and a student would be fined up to the full value of a book if it was spoiled or lost. I found that teaching school was more mentally than physically tiring, and I had little appetite for serious reading after school was over, but I could practice the clarinet.

About halfway through the summer session, the eighth grade civics class reached the chapter on Education. During the first month, I had tried to get comments from the students, especially in the upper grades, but with little success. When I encouraged comment on the subject of education, I discovered strong hidden emotions in the students. Hugh Lee Hunt made some bitter criticisms of education and was surprised when I thanked him and asked if any one else would like to speak, and several spoke. Fascinated by the opinions expressed, I had the class write a paper on the subject to be handed in the next day. It would take the place of the next examination, and I promised to give each student an A, regardless of what was said, for turning in at least two pages frankly giving his or her opinion about education. From the papers, which I still have, I noted that some of the opinions were close to those of Grandfather Johnston, who had grown up in this neighborhood. The authors, Hugh Lee and Ida Mae, are pictured in Figure 1.2(b). I quote from two of these letters with no change to the spelling or punctuation:

What I think about the school I go to,
by Hugh Lee Hunt

"I think school is a very good thing. it teaches us how to talk and how to be good citizen I think everone ought to know how to read and wright, etc. But if you are going to farm or something a high education

Hugh Lee Hunt

Ida Mae Tyson

Figure 1.2b. Students at Oak Grove School, Cherokee County, Georgia. Unclaimed in-school photographs taken by unknown traveling photographer, 1939.

is no good to you but if you are going to have a office job it is alright. But lots of people think they are better than others because they have a good education trying to be something when they know that they are NOTHING

But you never get back what you pay to get education. It is alright. Some young sprout get about half way quit and start teaching school think they have learned enough when they are about 18 or 19 and along there they think bigern a big man college students is nothing but blamed sissies. they are so dad-blamed biggety that they can't look at a poor person Most of them have got a cigar stuck in one corner of their mouth them that ain't is because it makes them sick

All the college pupils think they know more than anybody else I don't know anything and don't give a dad blame Most of them is likeing in the haid. there ain't nobody this side of heaven who can make me believe in education I have said a lot but that is just what I think. education is alright if you are agoing to teach school or something so it will do you some good. But if you are agoing to spend half you life getting education and then not use it will not do you any good and that is just crazy

> If you want education bad enough to spend half your life on it it is thier bussiness not mine. A teacher think he is doing something to put a problem on the board so you cant work it. he thinks he really knows something this old silly algebra will never has done anyone any good and never will just throwing your life away for nothing. that is just what I think about the school I go to the school building is just fine the worst thing is the teachers in it.
>
> What I think about education
> by Buford Sims
>
> I think education is a fine thang in some ways. It helps get a job. You can read wright and almost any thang you want to. But if you dont have a education you cant do nothing much. There is some studies that I dont like such as science and some other studies. If I was to tell you the truth I dont believe nothing they say.

This was the first student evaluation of my teaching. At family gatherings or in a group of intimate friends, a southern tradition is to relate a long elaborate joke that is a "true" story, where the speaker turns out to be the butt of the joke. In this tradition, I had memorized sections of Hugh Lee's paper, including authentic pronunciation and rhythm, and used it as the conclusion to my story about the Oak Grove School experience.

BAD NEWS

In my senior year, I was in sight of graduation, and I seriously dated a girl I liked. I thought of getting a desk job in the big Lockheed aircraft factory in Marietta, Georgia, near my home. I wanted to be independent, and I asked the doctor about my possibly getting married. He was considerate in how he said things, but he pointed out what problems I would cause to have a child or two and leave orphans in a few years. I had never thought about that. Anyhow, he said I couldn't stand the exertion of having sex "in the way that might produce a baby," and should only count on "hand or head" sex. He said I should send my fiancee in, and he "would explain to her what to expect." I said to myself I would never ask any girl to make a deal like that. However, after haunting me for years afterwards, the doctors words were found not to be true.

* * *

Occasionally, during college years, I attempted to write a novel or even a short story, but I could tell it was not good, and by my junior year in college I felt I could not have quick success as an author.

In the spring quarter of my junior year, I audited Professor Mose Harvey's Modern European History course. Instead of working through the textbook, which covered the period 1900 to 1932, Professor Harvey talked about the history of the previous week in Europe; Nazi Germany had knocked out Poland with their Blitzkrieg, captured Denmark and Norway, and launched Panzer divisions toward France, isolating an entire British army. In World War I there had been fake atrocity stories about the savage Huns, and many Emory students in 1939–40 were skeptical about the new round of atrocity stories and skeptical of Harvey's intensity. Perceiving this antagonism, Harvey recommended that the students read Hitler's book *Mein Kampf*, which I did. I was frightened and agreed with Harvey that Hitler sincerely intended to carry out every one of his terrible resolves. When other students still expressed their doubts, Professor Harvey became emotional and said words to the effect:

> Germany is about to knock out France, knock out Russia next, and then turn on Britain. If the United States is fast enough, we will go to war to save Britain. If we are not fast enough, we will find ourselves alone fighting the German empire with all the wealth of Europe behind it. You are free to be skeptical now, but every one of you will either join the armed forces or be drafted within a year or two. Think about that.

I thought about that. I looked around the room at my fellow students. "Some of them would look pretty funny in uniform. But what about me? What could I do?" (Much of this paragraph is based on and some is directly quoted from: Harold Johnston, *Journal of Physical Chemistry* **A105**, 1388–1390 (2001)).

The idea came as a surprise: I could change my major. I took out the Emory catalog to see if I could fulfill the requirements to enter medical school after one more year at Emory. It could be done, but the full medical education would take the next year of college, four years in medical school, and two years as an intern. I would be done seven years from now; "No, I don't have that much time." I consulted a friend and chemistry major, Nat Robertson, about switching to chemistry. As part of my science requirement, I had taken a year of freshman chemistry. Nat showed me that if I took five courses in summer school that year and four courses each quarter the next year, I could graduate in chemistry. I received special permission to take five courses in summer school, followed Nat's plan, and graduated at age twenty with a major in chemistry in June 1941.

CALIFORNIA INSTITUTE OF TECHNOLOGY CALTECH

Professor J. Sam Guy, Head of the chemistry department at Emory University, Atlanta, Georgia admired the California Institute of Technology (Caltech). It was a small institute with large faculty-student ratio. Unlike Caltech, the chemistry department at Emory was not carrying out, or much aware of, the major new scientific developments. Professor Guy wished he could bring some of the qualities of Caltech to Emory. Professor Guy advised me to apply for graduate school instead of trying to get a job after graduation. He said I should "shoot for the best" and apply to Caltech, but also apply elsewhere, including to two southern universities, where I had a better chance of being accepted. I am sure he sent Caltech a letter of recommendation that exaggerated my capabililties and led to the following surprising and delightful letter:

Mr. Harold S. Johnston April 1, 1941
Sigma Alpha Epsilon House
Emory University, Georgia

Dear Mr. Johnston:

I take pleasure in informing you that the Institute offers you an appointment as Graduate Assistant in Chemistry for the ten-month period commencing September 20, 1941. The stipend is five hundred and ten dollars to be paid in the following manner:

The Chemistry Division will pay to the Institute $360.00 covering your tuition charges for one year; the remaining $150.00 will be given to you in ten equal cash payments, the first to be made on September 30, 1941. You will be free from laboratory fees but may be charged for actual breakage and for twenty percent of the supplies used; and you will be asked to make a deposit of $15 to cover these charges.

The appointment as Graduate Assistant in Chemistry carries the following obligations:

(1) Teaching assistance in some branch of chemistry requiring a total expenditure of about ten to fifteen hours per week.

(2) Intensive devotion, during the academic year, of not less than one-third of the working time to scientific research, with graduate study occupying the rest of the time, and including preparation for qualifying examinations for Ph.D. candidacy; and continuance of research, full time, from the close of the academic year until July 20, 1942. Graduate credit is obtainable for this research. The research will be in a field to be agreed upon by you and the staff member with whom you work.

In accordance with an action taken by the Association of American Universities you are allowed until April 15 before final acceptance. We are, however, very desirous of knowing your answer since we shall wish to offer the appointment elsewhere in case you decline. We therefore urge that you inform us as soon as you have reached a decision.

Will you kindly acknowledge receipt of this letter.

Sincerely yours,
Roscoe G. Dickinson, Chairman
Graduate Committee of the Division of Chemistry and Chemical Engineering

---

The next day I received a letter from the University of Virginia offering me admission to graduate study in chemistry and a teaching assistantship for three years, which included tuition, full room and board in a dormitory, and a small cash stipend. Considering the burden of the depression on my family, I should have chosen the generous offer from Virginia, but Professor Guy had said I should "shoot for the best", and he said Caltech was the best. I discussed this dilemma with Professor Guy. He let me borrow the recently arrived March 17, 1941 issue of *LIFE* Magazine, which included a photographic essay of Caltech.[#] He said I should study this article and come back and talk to him before I mailed out my choice.

The *LIFE* article presented pictures and short descriptions of undergraduate and graduate students as well as the founding fathers and some of the great scientists at Caltech. Like Emory, Caltech was all male and operated on the honor system. Teachers handed out examinations and left the room; students were honor bound not to cheat and were honor bound to report any cheating that they observed.

Over the two decades after 1910, three scientists from great institutions in the eastern US and the brilliant scholars they brought in transformed the local Throop Polytechnic Institute, a school of trades like carpentry and plumbing, into the world-class California Institute of Technology:

Astronomer George Ellery Hale (1868–1938), educated at MIT, founded Mt. Wilson observatory, 1904, and designed the two-hundred inch (five m) reflecting telescope for Mt. Palomar observatory.

[#]Cal Tech: Creative Research Trains Scientists and Engineers, *LIFE*, March 17, 1941. Reprinted in spring of 2001 with added comments by Caltech Archives:
http://archives.caltech.edu/life_article/life_page_2.html

Chemist Arthur A. Noyes (1866–1936), educated at Harvard, was distinguished professor of inorganic chemistry at MIT and professor of chemistry at Caltech 1919–1936. Noyes and Hale laid down the policy of undergraduate and graduate student education.

Physicist Robert Andrews Millikan (1868–1953), educated at Columbia University, was Professor of Physics at University of Chicago, 1910, Director of Norman Bridge Laboratory of Physics at Caltech, 1921, and winner of the Nobel Prize in Physics, 1923. He was the leader in recruiting top-quality scientists and engineers to come to Caltech as faculty members.

Caltech had three Nobel Prize winners in 1941: Robert A. Millikan and Carl D. Anderson in physics, and Thomas Hunt Morgan in biology.

The *LIFE* article gave pictures and examples of high technology at Caltech in five fields:

(a) Chemistry: X-ray diffraction quantitatively reveals elaborate architecture of atoms in crystals.
(b) Physics: Nobel Prize winner, Prof. Carl Anderson's 2,000,000-volt electrostatic generator was being used to probe the structure of atomic nuclei.
(c) Biology: Prof. Alfred Sturtevant's fruit flies in jars pedigreed through 280 generations were giving new insight into genetics.
(d) Astronomy: Work was actively underway on the Caltech campus grinding the final curved surface of the 200 inch telescope mirror.
(e) Engineering: Caltech's 15-ft (4.6 m) triple-blade propeller was producing wind in a doughnut shaped wind tunnel at 200 miles (320 km) per hour. It could take aircraft models up to nine feet (2.7 m) wide. This wind tunnel had been important in the location and growth of the aircraft industry in southern California.

There were about 460 graduate students and professional research scholars. Caltech limited the number of undergraduate students to 640, and students were required to take two years of English, as well as courses in mathematics, physics, and chemistry. Pictures showed some students playing poker and others playing three-dimensional tick-tack-toe.

Everything about it glowed with charm and challenge. I strongly wanted to go to Caltech.

I felt reluctant to ask for so much money from my family, but I presented the advantages and disadvantages of each alternative to my parents. My father, mother, and I discussed all aspects of the question, and my mother said to postpone the decision until supper that night. By then she had telephoned my Atlanta doctor, found that Virginia had a much higher incidence of rheumatic fever than Georgia, and that this disease was extremely rare in southern California. My greatest immediate danger was to have another case of rheumatic fever, since one attack predisposes for subsequent attacks. So, my parents decided and I was delighted that I should go to Caltech.

Someone in Atlanta, who heard that I had been accepted for graduate school at Caltech, offered to bet my older brother, Smith, Jr., a hundred dollars that I would flunk out by Christmas, because, he said, a backward school such as Emory University could not prepare anyone to compete at that level. Smith was ready to take his bet, but our mother would not allow it, saying, "It would put too much pressure on him."

* * *

Professor Bill Jones, my physical chemistry teacher, was going to be on sabbatical leave at Caltech the next year, and in late August I rode out to California with him and his wife, stopping to see many mountains, canyons, and motels. We went through Oklahoma City to Colorado Springs, north to Yellowstone National Park, south to Zion National Monument, and much further south to cross the mountains into the Los Angeles basin. The picture postcards had shown clear deep-blue skies, shiny green leaves, and bright orange oranges, but as we went through the pass and began to go downhill, I was disappointed to note the gray-blue sky and poor long range visibility from a veil of smoke. When we reached Pasadena, I noted a gray haze that reduced the clarity of trees and buildings that were more than about fifty yards (46 m) away.

Figure 1.3 shows the symmetrical western entrance to Caltech focused on the distant dome of Throop Hall, the administration building, and gives a view of the east entrance of the Gates Chemical Laboratory with its elaborate decorations. The picture of the Gates Laboratory reveals a rope hanging from the attic level down to the entrance porch. This rope does not indicate a desperate hanging by an overworked student; rope climbing contests up the sides of buildings provided one of Caltech's favorite sporting events.[#]

[#]Cal Tech: Creative Research Trains Scientists and Engineers, *LIFE*, March 17, 1941. Reprinted in spring of 2001 with added comments by Caltech Archives: http://archives.caltech.edu/life_article/life_page_2.html

Western entrance to Caltech Campus, Throop Hall in Center Background, 1941. Unknown photographer gave this picture to author in 1942.

Figure 1.3. Gates Chemical Laboratory, California Institute of Technology, 1941. From family picture album, unknown photographer.

In the year 2000, Caltech issued to its alumni, "An Architectural Profile," which included pictures of some of its buildings, close-up pictures of some of their ornamental detail, and descriptive text which described the campus as carefully planned with "an eclectic mix of Mediterranean architecture recalling Andalusian villages and Florentine villas, featuring courtyards and patios."

* * *

Upon arriving at Caltech a few days before classes started, I reported to Professor Dickinson (Figure 1.4), and he advised me on what to do. He recommended the courses I should take. During the first two weeks, I asked to join his research group. He agreed to be my research advisor, and suggested that I work with Richard Noyes, a second year graduate student, to become familiar with the facilities and equipment. Because I had taken so little mathematics and science as an undergraduate, he said I should expect to take four years, not the usual three years, to get a Ph.D. degree. Professor Dickinson was aware of my health situation.

I entered Caltech on September 20, 1941, three weeks before my twenty-first birthday. I took *Advanced Inorganic Chemistry* from Professor Don Yost, *Thermodynamics and Physical Chemistry* from Professor S.J. Bates, and *The Nature of the Chemical Bond* from Professor Linus Pauling. Professor Bates was old. He would come to class, assign the problems for the next class meeting, take up the homework assigned at the previous meeting, and ask if anyone had any difficulties with the problems. If a student had a question, the professor would mumble to the blackboard with his back to the students, erase things and do it again differently, but only rarely did that clarify the question. Fortunately for me, my teaching duties were to grade papers in Professor Bates' undergraduate course in *Thermodynamics and Physical Chemistry*, which used the same textbook, covered the same topics as the graduate course, but moved along at a faster pace. I quickly learned which undergraduate students (Malcolm Mason and Shelton Steinle) got all the problems right; and so I learned how to do the problems, and could then grade the other undergraduates' papers—and pass the graduate-level course.

Professor Don Yost was short, slender, and elderly, but very much alive and with a subtle sense of humor. In a slightly halting manner of speech, he described inorganic chemicals in terms familiar to me, such as, color, method of preparation, reaction rates, and important reactions, and also in terms I had never heard of before, such as residual entropy, bond lengths

Figure 1.4. Professor Roscoe G. Dickinson, about 1940. Photograph provided by courtesy of California Institute of Technology Archives.

and angles, spectroscopic properties, and quantum mechanical states. It was easy to memorize it all, but I had to look up several things in other books to begin to understand the new class of properties.

Professor Linus Pauling, trim and athletic, walked briskly into the classroom with a big smile on his face. Class attendees included first-year graduate students who took the course for credit, advanced graduate students who audited the course for the pleasure of hearing it again, and visiting professors who audited the course to learn the material. With no notes,

Pauling talked throughout the period, except when someone asked a question, which he always answered clearly and completely. Each syllable in every word was distinctly and separately pronounced in an elegant voice. Some things he wrote on the blackboard, and some he wrote in the air with his hands and body. He was a skilled actor and put on a lively show. For me the trouble was that I did not understand what Pauling was talking about. I furiously took notes and studied them at night, but all of the material was far above my head. I took the entire four day Thanksgiving holiday weekend to do nothing but study Pauling's course with aid from Pauling's book, *Quantum Mechanics* and my lengthy class notes. By the end of that weekend, I had finally made some sense of the material and begun to understand it. There was only one examination, the final. If the final had been given before Thanksgiving, I would have flunked and gone back to Georgia hoping to find a job. By the time the final exam was given in December, I had understood enough to get a B in the course. With the grade of A in each of the other two courses, I had survived.

* * *

The Athenaeum was a luxurious faculty club, which admitted graduate students who associated across disciplinary lines and mixed with world famous faculty members. The Athenaeum was built in Mediterranean style with stone, stucco, red tile, much ornamentation, and marble on the ground-level floor and staircase. The main room appeared to be almost as large as a tennis court, most of the floor was covered with a huge single oriental rug, and the high walls were paneled with polished mahogany. At the Athenaeum, every male diner was required to wear a coat and tie.

On one occasion, at a table in the Athenaeum dining room, we were talking about classical music, records, record players, and speakers. I mentioned that I had a fairly good system, but that it was so bulky I had to leave it in Georgia. I said I hoped to buy another system out here. Tony, an electrical engineer, who had built his own tuners and amplifiers, said he would build me a system if I would pay for the material. He invited me to come up to his room to help him in the last stages of putting together and testing my system, but when I got there, another electrical engineering student was already helping him, and all I got to do was to look on. Tony told the other student that he had tried something different, and that he was proud it had worked out well. Tony had an old Model A Ford, which ran fine, but the seats were worn out, all the way through.

With my system working well, we drove his car to where I bought a record player and a good cheap second-hand speaker.

* * *

Late, in the fall quarter, three young Air Force officers were eating lunch in the Athenaeum at a table with several graduate students. They had stopped off on their way to Honolulu. One of the officers knew one of the students, and there was pleasant banter around the table. One officer stated as a matter of fact that we would be at war before Christmas 1941. The others of that table paid little attention to this remark, but I was struck by it and wrote a letter home quoting the officer.

* * *

My oldest uncle, Andy Johnston, and his wife Aunt Minnie lived in west Los Angeles. I had to study so hard during the fall quarter that I had not been able to visit them. After the final examinations were over, I visited them on Sunday, December 7, 1941. I took the streetcar from Pasadena to downtown Los Angeles and changed to another streetcar that went far out on Olympic Blvd. I got off and walked a few blocks to their house, arriving well before Sunday dinner. They had invited another couple in their age group, and Uncle Andy and the man had much in common and talked vigorously to each other. With help from the visiting woman, Aunt Minnie was cooking a generous dinner. Then Andy answered the telephone; the caller suggested that he turn on the radio. Soon we heard excited radio announcers telling about the assault on Pearl Harbor by Japanese aircraft. We listened without comment for perhaps a half hour, as the hot dinner in the other room cooled, and then Andy exploded in anger against the Japanese. He asked why we had let them do it. He anticipated that Japanese soldiers would come ashore to California.

* * *

I had passed my first quarter. I went to Professor Dickinson and asked to join his secret war project, but he told me that my job was to continue taking courses as a graduate student. Only after my grades were issued at the end of the winter quarter, three A's, Professor Dickinson called me in and asked, "Are you still interested in joining our war research project." I made a naively passionate declaration of: patriotism, hatred of the Nazis, and approval of joining the war project, ending with, "I want to do something useful." Careful not to laugh, the professor said, "Don't expect too much. You can't expect us to turn up any dead Nazis on this project.

> "We are guarding a bridge that may never be attacked; we hope it will not be. If it is not attacked, our work has succeeded."

As application for security clearance, I filled out a long questionnaire that included the dates of occupancy and address of every place I had lived. After a couple of weeks, I was granted a conditional CONFIDENTIAL security clearance, but my application for SECRET clearance had been delayed. Nevertheless Professor Dickinson took me on to his project, conditionally, and instructed that meanwhile I was not to read any SECRET document I came across in the lab. The delay in getting SECRET clearance dragged on for weeks, and to me it was a serious matter, because if I did not get it, I would not be allowed to stay on the project.

One Day, Hal

I stepped out from the old wooden Graduate Student Dormitory on the campus of the California Institute of Technology. I was dressed, as usual on the Caltech campus, in a long sleeve shirt, pants, jacket and tie, but with no hat. I stopped to admire the clear view of the San Gabriel mountains in the slanting early morning sunlight, and walked, with a slight self-satisfied swagger, to the south entrance of the Gates Chemical Laboratory. With effort, I swung open the tall heavy oak door, entered, and greeted the armed uniformed guard behind a desk in the hall. Ordinarily, workers and students merely showed their identification card to get past the guard, but for me this day was different. I said, "Good morning, Corporal. Here's my ID card, and I get to have my own key today. There should be a note from Professor Dickinson." The guard grunted, took the ID card, shuffled his papers, found the envelope, and said: "You gotta fill out this sheet. Do it in the library."

The chemistry department library was not up to the standards of a great institution like Caltech. It was a large, single, unsupervised room with book shelves on all sides except for a wide window facing a well-kept garden. The library had a swinging unlocked door accessible to anyone at any time, and its book collection was what was left over after professors and students had anonymously and permanently checked out everything of any interest. I sat down at a table and began to fill out the form.

REQUEST FOR KEY TO CLASSIFIED AREA

NAME (Print) HAL JOHNSTON **SOCIAL SECURITY NO.** [entered]

HEIGHT 5 ft. 10 in. **WEIGHT** 120 lb. **EYES** Blue **HAIR** Brown

SEX male **AGE.** 21 **BIRTHDAY** Oct. 11,1920

BIRTHPLACE Woodstock, Georgia **CITIZENSHIP** USA

MOTHER'S MAIDEN NAME Florine Dial **BIRTHPLACE** Woodstock, Ga.

FATHER'S NAME Smith Lemon Johnston, Sr. **BIRTHPLACE** Woodstock, Ga.

FACULTY___ **STUDENT**___ **EMPLOYEE** X , **OTHER**___

JOB DESCRIPTION Research Assistant

EMPLOYING AGENCY National Defense Research Committee (NDRC)

SECURITY CLASSIFICATION Confidential

SUPERVISOR Professor Roscoe Dickinson

PROJECT ROOMS FOR WHICH KEY IS REQUESTED 65 Crellin and 121 Gates

LOCAL RESIDENCE #15 Graduate student dormitory, Caltech

SIGNATURE [I signed] **DATE** [Early May, date entered]

KEY TO BE RETURNED AT THE END OF EACH DAY.

In the lower right hand corner of the sheet there was a blank square space for a THUMBPRINT. Upon checking Employee, I silently noted, "Six weeks ago, I was only a graduate student." Upon writing Confidential, I silently said, "I expect to have SECRET clearance soon." I returned the sheet to the guard. The guard slowly checked the entries, occasionally looking up to see if the descriptions were correct, and then took out his ink pad, saying, "Gimme your right thumb." I presented my right hand, its baby smooth skin announced the fact that I had almost never done any hard work. For eight years the hardest work those fingers had done was to play the clarinet. The thumbprint was taken, and I wiped off the excess ink with the offered square of brown paper. The guard pulled out another sheet, and said, "Sign here, in and out. Bring the key back."

KEY RECEIPT TIME 6:30 A.M. SIGNATURE [I signed]

KEY RETURN TIME ___ SIGNATURE ___

I thanked the guard and walked downstairs to the sub-basement of Crellin Hall. An insulated chest containing bricks of Dry Ice stood by the elevator, next to a stack of cardboard boxes. I selected a box, put on the provided insulated glove, and half filled the box with bricks of Dry Ice. I took the elevator up one floor to the basement of Crellin and walked down the hall to room 65. A large sign said: "RESTRICTED AREA, SECURITY CLEARANCE REQUIRED, UNAUTHORIZED ENTRY SUBJECT TO CRIMINAL PROSECUTION IN FEDERAL COURT" I read the sign and grinned with pride. This was my fifth week on the NDRC project and the first day they had let me have a key. I had come in early that day to do the Dry Ice and other routine jobs, so that the others could all do a full day's work when they got there at about eight a.m; this was a big day, the army was going to come in and pick up the special stuff that our group had made. I unlocked the door with the key, and opened the door, which had a spring that kept it from being left open. I pushed the box of Dry Ice in with one foot and entered the room.

The room was shaped like the letter L, upside down, wrapped around a small square room. The entrance was at the end of the long arm of the L, and the short arm of the L was to the left at the far side of the room. A long laboratory bench stretched along three quarters of the right side of the room, and it contained a large fume hood. A fume hood is like a kitchen exhaust fan, except it is six feet (1.8.m) wide, sealed on all sides except the front, which is covered with a window of explosion-resistant glass. The window could be raised or lowered against balancing weights. A second fume hood was on the left side of the room at the short end of the L. Fume hoods are common in chemical laboratories. These carried air from the room up a chimney pipe. I was told that Professor Dickinson was a "safety freak" and had demanded and obtained new fume-hood motors with twice the pumping capacity of the old ones; he also had the exhaust chimney extended to well above the roof. The two fume hoods took air out of the room, to be replaced by filtered fresh air from an input located in the middle of the back wall.

Each hood had a sign, "HAZARDOUS VOLATILE MATERIALS. ONLY QUALIFIED PERSONNEL SHOULD OPEN THIS WINDOW". I silently said to the fume hood, "Hey, kid, I'm qualified. Let's see how cold your insides are." I opened the window to the hood, which contained a few small stainless steel gas tanks with shut-off valves, two six-liter glass bulbs with sealed-on stopcocks, a small rack of sealed glass bulbs with liquids inside, and a row of Dewar flasks which were large open-mouth double-walled thermos bottles. These flasks were large enough that the Dry Ice they contained needed to be renewed only once a day. Each Dewar flask contained a sealed glass cylindrical bulb, which had a long fat neck, tightly clamped to a fixed bar. Each bulb contained a solid or liquid inside.

Though the bulbs' contents were gaseous at room temperature, they were liquid or solid at Dry Ice temperature. A piece of rumpled cheesecloth was wrapped repeatedly around the top of each Dewar flask to inhibit moist laboratory air from coming in contact with the cold Dry Ice inside the flask. As the first thing I did each day, I added Dry Ice to each flask and replaced the cheesecloth by a dry one. Once a week each bulb had to be removed from its cold bath and placed in a freshly loaded one, because of the build-up of ice in the cold slurry. As the youngest member of the group, I had the job of re-supplying the Dry Ice every day and re-loading the flasks on Fridays. Each glass bulb, Dewar flask, and bulb stem had a label to identify its contents.

I made a quick survey of each bulb in the two hoods to see if there was any problem. The first bulb contained phosgene, Cl_2CO; it was our calibrating gas. It was used as a war gas in World War I. As I understood it, phosgene dissolves slowly in water and passes through the nose and throat with only moderate irritation, but it reacts in the lungs to form two units of hydrochloric acid, HCl, which corrodes the lungs and can lead to death within twenty-four hours. The second bulb contained chloro-fluoro phosgene, ClFCO, which was said to be much more toxic than regular phosgene. It breaks down in the lungs to form one unit of hydrochloric acid and one unit of hydrofluoric acid, HF, which is a weaker acid than hydrochloric acid, but is more persistent and does more physiological damage. The third bulb contained sulfuryl chloride fluoride, $ClFSO_2$. It has the potential of breaking down in the lungs to give one unit of sulfuric acid plus the acids carried by chloro-fluoro phosgene. The material in the fourth and fifth bulbs had arrived recently and came with special warnings. These gases quickly shut the pupils of the eyes, paralyze the lungs, and were fast killers. Our group nick-named them "di-methyl poof" and "di-isopropyl poof," with the nick-name "poofs" standing for fluoro-phosphates (Later our army captured similar substances in Germany, and they are called nerve gases. The fluorophosphates in German nerve gases were much more toxic than those we worked with). The fluorophosphates had not yet been run through the laboratory tests. In the past, other new chemicals not now on hand had been tested by Dickinson's group and so we would test these new chemicals too.

Ref. *Phosgene Toxicity*, Published by Institution of Chemical Engineers, Copyright 1993. Warwick Printing Company Limited, Theatre Street, Warwick CV 34 4DR, UK.

In other laboratories, toxicologists and chemists worked together to identify and synthesize compounds known or suspected to be poisonous, volatile, and stable enough to be used as a war gas. In yet other laboratories, animals were exposed to various amounts of these gases to test their toxicity. My associates did not know where these laboratories were. Even though my associates had SECRET clearance, they had no "need to know," and the information was kept compartmented. In the laboratory, we had four metal cans filled with different charcoals, simply labeled: CHAR I, CHAR II ... The details of these charcoals were SECRET. Before I joined the group, they had built an array of gas tanks, glass and metal tubes, large glass bulbs, vacuum pumps, a high temperature furnace, electrical conductivity cells, detectors, and Esterline-Angus recording meters. Initially, the system looked like a Rube Goldberg device, but as I came to understand its workings I was amazed at the scientific sophistication that went into some of its components. The groups major job was to determine for how long a time a charcoal would give protection against any new poison gas the enemy might come up with. To do this, we flowed a stream of air and gas through a bed of charcoal and measured the time it took for the gas to break through the bed. We repeated the tests with different amounts of the gas in air, at different temperatures, and with different charcoals. We tested the different charcoals according to instructions from Professor Dickinson. This system occupied the desk space beyond the fume hood on the right rear of the room.

The chemical warfare service of the War and Navy Departments carried out experiments of this sort by passing air and gas through a bed of charcoal, collecting about a dozen samples of the gas-air mixture after it went through the charcoal, and submitting each sample to a tedious chemical analysis called "titration." Dickinson's method was at least a hundred times faster and gave much more information than the traditional army method.

* * *

My immediate supervisor was John Otvos, a skilled and patient teacher, even to one as green as I was. John was a third-year graduate student, or would have been if he had not joined this NDRC project. John had first come to my attention the previous fall. A group of eight graduate students

General ref. in this chapter: Marschall Sittig, *Handbook of Toxic and Hazardous Chemicals*, Noyes Publications, Park Ridge, New Jersey, 1981.

were seated around the dinner table in the Caltech Athenaeum. One student, who apparently had just discovered Sigmund Freud, dominated the conversation, covering ground that most of his listeners had traversed years before. He explained that a man who loves his mother has an "Oedipus Complex" and a woman who loves her father has an "Electra Complex." John Otvos succeeded in entering the discussion. He asked in a sincere gentle voice as if he really needed the information: "Tell me. What name would they call me? I love both my mother and my father." I have admired John ever since and was glad to find him in Professor Dickinson's group, upon joining it.

On the first day when I joined the group, John showed me the various features of the project. When we came to the second hood, he lifted a small sealed glass tube out of its Dry Ice slush; it was about half full of a clear liquid. He said in a matter-of-fact voice: "That is sulfur decafluoride, S_2F_{10}, (abbreviated here as S-10) sometimes we call it Stoff-Stoff, pronounced in German. It is a close relative to sulfur hexafluoride, SF_6, which is about the most unreactive compound known to chemistry. S-10 is colorless, odorless, and four times as poisonous as phosgene. One breath of it and you die within a day, drowning in your own water and blood, and there is nothing anybody can do about it." I felt an almost pleasant tingle of fear. John explained that they had made some selenium decafluoride, selenium being in the same chemical family as sulfur, but they no longer had any of it on hand. It was thought to be even more poisonous than Stoff-Stoff. They needed a code name for it. The name, selenium, is based on a Greek stem meaning moon. They decided to call it Moon-Stoff. John and his parents came from Hungary, and John knew and spoke Hungarian as his first language. To further complicate the code name, John had named selenium decafluoride Hold-Stoff, since *hold* is the Hungarian word for moon.

Arthur Stosick had finished his Ph.D. in chemistry, and he had stayed on as a postdoctoral fellow to set up Dickinson's project. Art's skills and knowledge had no bounds, so far as I could tell. He put together a few vacuum tubes and other electrical supplies to guide the operation of many tasks in the laboratory. The group did the chemical analyses by ingenious circuits that Art had designed and built, and recorded the final output of a charcoal test on an Esterline Angus chart recorder. Art was married, lived off campus, and was too busy for small talk.

I thought of Professor Dickinson as elderly, thinking "he must be over fifty." He was forty-eight. Professor Dickinson was short, his full head of black hair and his small black mustache were lightly streaked with white hairs. He wore horn-rim glasses, had a quick grin, and sometimes called me "young fellow." My birthday was on the same day and year as Professor Dickinson's son, Robert. The professor was Dean of the Caltech Graduate Division, did actual work in the laboratory, but was frequently called out on Institute and other business.

John Otvos, and I are shown in Figure 1.5. I never found a picture of Art Stosick.

* * *

The second task of this project was to manufacture enough S-10 for the army to test it in a small bomb to see if it would stand up to such treatment. We did this work in 121 Gates. Both for 65 Crellin and for 121 Gates, Professor Dickinson gave me several sheets of paper that spelled out the hazards of the gases and safety procedures to be followed with each apparatus, and he insisted that I study them. After that first day, I approached the S-10 hood with especial caution, although some of the other chemicals were equally or more poisonous; the "colorless odorless" part of it made it different. At least, CHAR I stopped S-10 to a satisfactory degree.

There had been no facility in the country set up to make a large amount of S-10. Professor Dickinson consulted the famous ingenious inorganic chemist at Caltech, Professor Don Yost, who knew a better way to make S-10 than that used by the army to get their sample. Yost's reaction cell had high temperature on one end and low temperature on the other end, which gave a relatively high yield of S-10. Yost had on hand an apparatus which was a giant fluorine maker that could be used. Dickinson got approval from NDRC and the Army Chemical Corps to proceed, and was granted extra funds to set up the apparatus and to hire a fourth person, and that was how I got my job. It was a job that I had asked Professor Dickinson for, in the previous December, when I received my grades for the fall quarter at Caltech.

* * *

The large fluorine generator was set up in Room 121 Gates. John Otvos and Arthur Stosick developed the procedure, and they manufactured some S-10 every day. They ran the fluorinator every morning as I watched during the first week; in the afternoon John and Art worked together doing the charcoal work, as I looked on. Later, John and I took over both jobs, releasing Art to do other work. Later still, John let me do almost every operation

Hal Johnston. Studio photograph, 1941.

Figure 1.5. John Otvos. Photograph by Army photographer in Panama, 1944.

in the S-10 preparation, under his supervision, and in the afternoons I became a specialist on one-half of the charcoal system. It had taken more than a month to fill a large bulb in the second hood of Room 65, with the amount of S-10 the army needed. The dangerous job of transferring the S-10 from the small bulb of each day's work to the large container in the second hood was done only by Professor Dickinson, or sometimes by Dr. Art Stosick. On this special day, the large glass container was filled to the desired mark with S-10, and a group from the army was going to pick up the sample in a container they had fabricated

S-10 does not dissolve in water, or strong acid, or strong alkali. We had to pass it down a hot monel tube to break it down so we could read its electrical resistance in water, the method we used to detect when a gas "broke through" the bed of charcoal. I silently consulted with myself: "How can such an inert molecule as this do so much damage in the lungs?" The molecular structure is $F_5S\text{-}SF_5$, two identical twins joined together. Considering new things I had learned at Caltech, I came up with a theory. "DISPROPORTIONATION, a specific technical word, one goes up and one goes down. That's it. DISS-pruh-PORE-shuh-NAY-shun, a beautiful word, a line of poetry in one word. Somehow in the lungs a biological reaction causes S_2F_{10} to disproportionate into SF_6 and SF_4. SF_6 is stable, virtually nothing reacts with it, it lives almost forever. SF_4 is a reactive beast, it tears water apart to make four molecules of hydrofluoric acid and one molecule of sulfurous acid; its life is short. 'Greater love hath no molecule than this, that he lay down his chemical potential for his friend.' " I loved to play around with famous quotations.

I snapped out of this daydream, afraid I had wasted too much time, and looked up at the clock. There was still plenty of time to smash up the Dry Ice and distribute it to the Dewars before the rest of them came in. I rushed over to the box of Dry Ice, got out the large cast iron mortar and pestle, put a brick of Dry Ice in the mortar, took the iron scoop in my left hand, took the heavy iron pestle in my right hand, kneeled in front of the mortar to get good access to it, and began to smash up the cold bricks with mighty strokes of the pestle. When one brick was turned to powder, I scooped it out and put it in a box, and grabbed another brick. Soon I started to make a game out of the job, clanged the iron scoop against the mortar as I lifted the pestle, and brought it down with a thump; clang-clang — thump, clang-clang — thump, to the rhythm of the anvil chorus of *Il Trovotore*.

This juvenile exuberance was cut short by the return, for the first time in a month, of the pain. The pain hit like a stab in the left shoulder. I perceived the branching of nerves as the pain branched down to the left elbow and up to the chin. The pain vibrated a few times and then vanished. I let the scoop and pestle drop and sat on the floor for a moment. I picked up the scoop and pestle, and with slow deliberate actions finished the job of crushing the Dry Ice. I took the box of crushed Dry Ice to the first hood, unwound the stiff, icy cheesecloth from the phosgene container and dropped it on the floor, picked a ring of ice off the rim of the Dewar with my fingers and off the side of the cylinder of phosgene, filled the Dewar with crushed Dry Ice using a large plastic spoon, and wrapped a fresh dry cheesecloth around the top of the Dewar. Step by step, I did the same for the other samples in the hood. I added Dry Ice to the Stoff-Stoff in hood number two, the Stoff-Stoff the army was going to pick up that day. I put the wet cheesecloths in a cardboard box to be delivered later to the janitor, who would have them washed and dried. The whole operation was finished before anyone else arrived. John came well before eight, Art and Professor Dickinson a few minutes later. As each arrived, each cordially greeted the others who had come in earlier.

Dickinson quickly summarized the work of the morning: He and Art would transfer the S-10 in the hood to the army's container when it arrived. John and I were to do charcoal work on the newly arrived samples. John should begin to set up now, but Dickinson said I should come with him, since a member of the army team wanted to talk with me. The army group was there already and would deliver their container for the S-10. After the army group leaves, Dickinson and Art would close down Room 121, making safe all hazardous features, but leaving the apparatus ready for further work if the army should need some more S-10.

* * *

Professor Dickinson and I went to a small conference room reserved for this interview, and I was introduced to Captain Philip Everett, who was there in an army uniform with the Chemical Warfare Insignia: "How do you do, Mr. Johnston."

"How do you do, Captain Everett, sir?"

The professor left. The Captain was a big but not fat man, about forty years old, and he spoke in a formal manner. "Mr. Johnston," he said, "I am a medical officer in the Chemical Warfare Service. Questions have been raised in connection with your application for SECRET clearance and concerning your suitability to work safely on Dickinson's project. I am here to interview you concerning these and other questions." I was shocked at the idea that I might not be able to work safely in Dickinson's laboratory.

"In your application for clearance, there is a brief statement that you have some heart problem. Do you have heart trouble?"

"Yes, sir."

"I understand your southern upbringing, but as a civilian, you need not call me *sir*, and I would appreciate it if you don't. Now tell me about it."

I summarized what I understood about my heart problem, including my understanding about my limited life expectation.

He asked, "How often do you faint?"

"I have never fainted."

"Do you ever feel dizzy, light in the head?"

"No, sir. I mean, no."

He had me take off my shirt, and he did a standard examination. He asked more questions, and concluded, "Well, that finishes my examination. For NDRC and for the army, I will have to enter on the record your limitations because of the heart problem. It is for your protection as well as theirs. You can go, now, but be careful with those agents you are dealing with; any one of them could cause you to die very young."

"Oh, yes. We are very careful. Thank you very much. I'll go now." I returned to the laboratory.

* * *

The army group had brought in its glass flask attached to a carrying rack to receive the S-10, but it was too tall to fit in the hood, and so had to be filled by vacuum distillation from the gas supply inside the hood and out into the Army's container. Stosick clamped it to a heavy wooden desk outside the hood. The large glass cylindrical bulb was built into a frame and mounted about half a meter from the hood. The transfer would require liquid-air cooling of the external bulb. Art went to bring in a container of liquid air.

John and I carried out the charcoal tests for the newly arrived gases according to plan, but at slack intervals in our job, we watched the ongoing operation. Dickinson and Stosick had finished making the all-glass connection. They turned on the pumps to clear the apparatus of air. Stosick mounted a large-mouth Dewar flask containing liquid air under the Army's empty bulb. Slowly he turned the crank of an automobile jack to raise the liquid air up to and onto the bulb. Upon contact with the warm bulb, some cold liquid air vigorously boiled away and formed a cold fog, which settled to the floor, spread out, and evaporated. S-10 began to come over and condense out, which boiled off some more liquid air. Dickinson watched the S-10 in the hood and told Stosick at what rate to raise the liquid air. S-10 boils (just) above room temperature, so a vapor-pressure explosion could not occur. The whole process was over sooner than Dickinson had thought. The glass line between the hood and the army bulb was sealed off and disconnected; Dickinson was a master glass blower. Stosick slowly lowered the liquid air Dewar below the sample and removed it from the Army's rack. They then took off their glass-blowing goggles, removed the gas masks from their belt, sat down and smoked one cigarette after another.

The backboard of the table hid the bulb full of deadly S-10 from easy view. First John and then I walked around to the side to get a close look. Moist air contacted the exterior of the cold bulb and formed a layer of ice.

We had a local telephone in the laboratory, but to make an outside telephone call, Dickinson had a private line in his office nearby; "I'm going to telephone them to come pick it up. Art, why don't you take a walk outside and take it easy for a while," emphatically, "Everybody else stay away from it." A few minutes later, he came back, "They won't pick it up until about two PM."

* * *

Then the laboratory was calm, all were busy, and none was rushed. Stosick was fabricating some electronic device on the laboratory bench to the right of the entrance door. Dickinson sat at his desk to the left of the entrance door, busy with paper work. John and I began work on one of the "poofs" (organic fluorophosphates) at the center rear of the main laboratory. The Army's rack with its S-10 stood on the desk about half a meter from the hood and about three meters from John and me.

John worked the input side of the system on the laboratory bench and transferred some of the "poof" from its cold liquid state into a gas in a spherical twelve-liter glass bulb. He slowly filled the bulb with pure dry nitrogen from a large high pressure tank and proceeded to mix the "poof" and nitrogen in the flask. A constant flow of nitrogen gas went into the twelve-liter flask and out another tube. We could let the flow go directly to the chemical analysis system, or by rotating a stopcock, we could first flow the gas through a bed of charcoal and then through the chemical analysis system. The chemical analysis involved a high temperature furnace to break down the "poof," a counter-current gas-water flow to dissolve the decomposition products in water, and an electrical device that measured the resistance of the water solution, whose output went to a recording meter. For a time, the recording pen traced a straight line across the bottom of the flowing paper chart, and all John or I had to do for several minutes was to wait for the ink pen to rise on the chart paper. During one of these waiting intervals, I leaned back in a chair, put my feet up on another chair, and mimicked smoking a cigar: "What an easy life. When I graduate and get a job, I'm going to do my work with vacuum tubes and recording meters. No more burettes. This is fun." John laughed politely, Art glanced up and smiled, but nobody knew what Professor Dickinson thought, and I quickly got back to work before I found out.

* * *

Glancing over the equipment used for the charcoal tests, I recalled a day at the end of my second week on the project. A wide, strongly braced, shelf held a large, water-filled carboy about four feet (1.2 m) above the laboratory table. The carboy had a narrow neck at its top center and a narrow neck on its side near the bottom, and each neck was sealed by a one-hole rubber stopper with a glass tube through its center. The top-center glass tube extended about three centimeters below the rubber stopper, and above the stopper it held a filter that protected the distilled water from dust or other contamination as laboratory air passed into the carboy. The bottom glass tube was inserted into a long clean rubber tube that hung down almost to the table top. The rubber tube was clamped to prevent the flow of water until it was needed, and it was terminated with a glass capillary tube, which controlled the flow rate. A meter stick was taped on the side of the carboy to indicate the height of its water.

During my first two weeks at work, one of my tasks was to record the height of water at different times, while John and Art carried out a charcoal experiment. Toward the end of the second week, John showed me a calibration curve of water flowrate for different heights of water in the carboy, and, always an educator, suggested that I derive a formula for the flow rate through the capillary tube. That was an easy assignment: the pressure on the input side of the capillary would be the atmosphere pressing on the water surface plus the pressure of the water column between its height in the carboy and its height at the capillary. The emitted water would be at one-atmosphere pressure. Assuming the flowrate to be proportional to pressure difference, I had the desired formula. I showed my work to John, who agreed that it was correct. I then dared to say, "Why do you do it that way? If you insert a glass tube from the center stopper to the bottom of the carboy, the pressure across the capillary would always be the same, regardless of how much water was in the carboy." I scribbled out a diagram to illustrate the concept.

At first John said nothing, but then said simply, "That is right. Let's show it to Professor Dickinson."

Professor Dickinson saw the point at once, "What led you to figure this out?"

I gave more explanation than expected, "At my home in Woodstock, we keep some chickens, which have to be fed and watered. If you put out a tub of water, they quickly mess it up. At the store, you can buy a small shallow pan shaped like a four pointed star. It screws onto a half-gallon (two liters) glass jar. Fill the jar with water, screw on the star-shaped lid, and quickly turn the jar upside down. When the chickens drink from the star-shaped trough, a bubble of air goes in and a slug of water comes out. The water level in the drinking trough remains the same regardless of how much water is in the half-gallon jar."

As a genial joke, Professor Dickinson said, "John, why didn't you think of that?" John replied, "Well, I guess it's because I never was a chicken farmer." Professor Dickinson went to the glass-blowing bench, cut a glass tube of the proper length, fire-polished the ends, put it in the carboy, and it worked as expected. At this point Professor Dickinson complimented me for the idea, and I felt proud of myself but tried not to show it.

* * *

After receiving a telephone call late in the morning, Dickinson cheerfully told me that my SECRET clearance was approved.

At noon, we took an hour off for lunch. Art always went his own way. John had brought his lunch and went across the street to Tournament Park, to toss soft balls as practice. I had a quick lunch at the Greasy Spoon on the ground floor of the graduate student dormitory. I sat on the far stool at the counter, so that no one could sit next to me and disturb my thoughts: I had my security clearance. The uncertainty was over. This was my happy day. I boasted to myself, "I did it. I hit the jackpot twice, I got in at Caltech and now I really have this exciting, important war research."

I was categorized as 4F in the draft: "Mentally or physically unfit for service in the armed forces." I thought of myself as being highly patriotic in joining the war project after two years of planning. I said to myself a proud private boast, "I have defeated the draft system—in reverse."

Upon hindsight, it is clear that my real motive was that I didn't want to be left out.

* * *

As I looked at one of the female cooks in the kitchen alcove of the Greasy Spoon, I noted that she was what my mother would call a "colored girl," and what my father would call a Negro, where the capital letters indicate that the word was pronounced with respect. Southern white people had many spellings and pronunciations, such as: NeGro pronounced knee-grow with accent on both syllables, Negro with accent only on the first syllable, negro, nigra, nigger, and nigah. In my family the word "nigger" was strongly forbidden; it was worse than a swear word. My mother hired a young black woman to do substantial housework, cleaning, cooking, and baby sitting. With affection I recalled Minnie, who worked for them from when I was age two to eight. As a small boy, I had played with a Negro boy about my age, and we were friends. After entering the first grade of school, I rarely saw that friend again.

In the third grade, my best friend Frank owned a billy goat, and we had fun as the goat ran rapidly and erratically down the sidewalk pulling a wagon just big enough for two small boys. As we were playing at the edge of the woods one day, Frank picked up a broken pine limb, pulled off the brown needles and some twigs, and aimed it as if it were a gun. "Bang, bang, I killed one," Frank sang out.

"What did you kill?"

Frank replied triumphantly, "A nigger. I'm practicing for the war between the whites and the niggers. Bang, bang."

"Who said there is going to be a war?"

Frank said that soon that is just what is going to happen, and we should store up guns and bullets to be ready.

I asked my parents about such a war. They told me that such talk was nonsense and wicked. Emphatically they said that, no such war was going to happen.

Frank and I argued about it, and our friendship broke up.

Through the local church, I established a connection with the Methodist Episcopal Congo Mission, Belgian Congo, Africa, and exchanged several rounds of correspondence with Lolonga D., an African boy my age. Lolonga wrote, but not in English; a missionary there translated our letters, which I still have. We expressed Christian love for each other, and I regarded him as a good friend. Looking at the cook in 1942, I recalled a daydream from about 1935:

> Lolonga had come to the United States for a visit, including a stay in Woodstock. I met him at the train station and planned to have him stay in our home. Also, I planned to take Lolonga to school with me and

> introduce him to the teachers and fellow schoolmates. I suddenly realized what I was planning; "No! It won't work. He would not be allowed in my school." Growing up in a race-based caste system, I had always realized the "Darkies" were people, but "not like us." Now I said to myself, "But Lolonga is a person, like me, why shouldn't he be allowed?"

With this faint spark of enlightenment, I began to say things that made my parents cringe, and some of my classmates called me a "nigger lover," but so long as I remained in the South, my actions differed little from those of the average southerner. Upon entering Caltech, I had been surprised and disappointed that I had not seen a single African American student there.

* * *

After finishing lunch, I walked twice around the perimeter of Caltech, a short walk. I recalled a pleasant weekend with my father at a campout at Etowah Mills, where the river was dammed by a cement and rock wall about twelve feet (3.7 m) high to form a long narrow lake. The overflow of the dam was channeled to a large wheel on the outside of the wooden millhouse. The wheel had a number of wooden bins that filled with flowing water at the top of the wheel and discharged at the bottom, powering the machinery inside. My father was a Sunday School teacher and leader of the group of young men at this outing. I, about six or seven years old, was brought along, but not my older brother Smith nor my younger brother Dick. The young men spent the afternoon swimming and boating on the lake; one of the men perched me on his shoulders and bounced in shallow water, and my father took me along in a boat. I was the center of attention and enjoyed it. We cooked dinner over campfires. Wrapped in blankets and quilts, we all slept on the wooden floor of the mill.

* * *

I concluded the walk and headed back to the laboratory. The air was clean, a pleasant breeze blew from the west, the Caltech campus was beautiful, and I noted again that this was my happy day.

I returned early to the laboratory, by-passing undergraduate students as they lined up to get into their one o'clock class. John and I worked on the methyl fluorophosphate tests. The regular charcoal seemed satisfactorily to stop the first sample we tested, but we would have to do farther tests for several different flow rates, dilution, humidity, and other parameters. Dickinson and Stosick, talking together, got back somewhat before two o'clock. The two of them turned to the left to inspect the Stoff-Stoff.

I heard loud and then soft exclamations from Dickinson. "IT'S ALL GONE. Cracked. Hours ago, probably. I'll go call the army."

As they turned away from the left corner, one stricken pair looked at the other pair who were just catching on. Dickinson told John and me what had happened, "They didn't anneal their glass bulb. It cracked and leaked. It's all gone."

Fear grabbed my throat and guts, my heart raced and it included irregular thumps. An image flashed in my mind of John, saying, "colorless, odorless, and four times as poisonous as phosgene." Then I glanced quickly at the others. They showed no fear, and so I concentrated on not showing mine. Silently I thought, "This is how soldiers can jump out of the trench together and rush at the enemy machine gun."

Dickinson conjectured that it probably happened during lunch, because we didn't hear anything. Furthermore, he said the laboratory had excellent ventilation, air came in at the middle and went out the hoods, the S-10 was close to the hood, and it probably went up the stack. Dickinson told John and me to go up to Gates and make another batch of S-10, while he and Art would call the army, return to clean up the laboratory, and carry on the tests John and I had started.

I recalled a preacher who said he was asked what he would do if told that he would drop dead tomorrow; and he listed, one after the other, exactly those things he planned to do anyway. The preacher said one should live each day as if it were the last day. I silently noted that Professor Dickinson was advocating such a course for this day.

My thoughts continued along another path, Professor Dickinson said the ventilation system carried it all up the hood, but I didn't believe it. Dickinson smoked much of the time. He could blow elegant smoke rings into the air. At idle times I watched a smoke ring distort, fold, stretch out, and drift around the lab until it became so dilute that it could no longer be followed. From the trails of many smoke rings that I had watched, I knew that the laboratory air did not march from fresh-air input to the hoods. It swirled and swooped all over the lab. I silently noted that when the liquid-air fog formed on the table just in front of the hood, the cold fog fell to the floor, it didn't go directly into the hood. As we walked through the hall toward Gates, I avoided looking directly at John and hoped John was not looking too closely at me. We didn't speak. As we passed graduate student friends in the hall, we did not speak to them. By the time we reached 121 Gates, the alarming fear of instant death faded into a low-intensity sad sick feeling.

John excused himself to pick up some tools from the shop, and I stood outside the door to wait. From my recent reading on the effect of lung-attacking gases such as S-10, I understood that a heavy exposure would show symptoms immediately, but a moderate exposure could be fatal and wait as long as five hours to show the first symptoms, and there was no useful preliminary treatment to head off or modify the effects after an exposure. "Dickinson wants to keep us busy, until ... A bomb filling of S-10 had leaked into the lab; this is it; it has happened already." I knew that I would die that night, and my recollections became more serious.

Standing outside room 121 Gates waiting for John to return from the shop, I compared my reaction eight years ago to that just experienced, and I noted the similarity. I admitted that today I had been terribly frightened, but that within less than a minute I had gotten over it. Of course, fifteen hours was different from fifteen years—I had asked for it—but I didn't expect things to happen so fast. I was determined that nobody was going to see me being afraid.

* * *

121 Gates is a small square room. We went promptly to work, freely speaking to each other only as needed to operate the system. The fluorine maker was an awkward looking machine. It was mounted on strong steel legs about eighteen inches (0.5 m) above the floor, and stood just in front of the hood for a total height of about five feet (1.5 m). The lower two thirds was a large pot that contained sodium fluoride and hydrofluoric acid or HF. The upper third of the fluorine machine had a filter of sodium fluoride that removed HF from the fluorine gas, F_2, and a high/low temperature reactor. The job had three distinct stages. The pot had to be brought up to the correct moderately high temperature to melt the salt and acid mixture, and then a direct electrical current was applied to form fluorine. Argon gas flowed through the top of the pot and carried fluorine out the exit line. Next, the flowing fluorine was diverted to the reaction cell in a monel tube furnace, which was heated on one end and cooled on the other end. The fluorine reacted with sulfur to make inert harmless SF_6 and highly poisonous S_2F_{10}. The army made its S_2F_{10} in a high-temperature system; Professor Yost's method of hot and cold conditions gave much larger yields of S_2F_{10}. Finally, the freshly prepared S-10 was slowly vacuum distilled to remove trapped HF and SF_6. Detectors and meters that Art had designed and built guided some of these operations. The full process took three hours to complete, and then the system had to be cleaned up.

An exterior electrical heater melted the mixture in the pot. John slowly turned up the heating current, closely watching three meters whilst writing it all down. When it reached the correct temperature, I turned on a direct current through the melt to form elemental fluorine, F_2, and I read one meter. The formation of fluorine depleted the amount of HF in the melt, which tended to cause the melt to solidify, but I continually added new HF gas to maintain the fluidity of the mixture. It was a tricky job to add the HF at just the right rate. On the laboratory bench, there were two large open jars of a paste to be used as soon as possible if any HF got on a person's skin. HF causes deep ulcers that take a long time to heal. Safety called for rubber gloves and face shields, but operation of the machine called for fast free fingers and a clear sight of what went on. Mostly we wore goggles as eye protection, but took our chances with bare, unshielded, fingers. The machine popped and bumped from its belly. Occasionally, these tiny explosions spit specks of hot sodium fluoride, smoking with HF, out the open exit of the fluorine maker into the hood. I cleaned up these specks during the last phase of the job. The current was maintained steady until these noises calmed down, and then I raised the current another notch. John tested to see if the machine was making fluorine. He picked up a small cotton ball with long metal tweezers, moistened it lightly on one side with alcohol, and held it up to the exit. When the cotton ball caught on fire, it meant that fluorine was coming out. Fluorine gas, F_2, reacts with almost everything; it is the most reactive element. John weighed out the sulfur and put it in a copper boat and into the monel tube furnace, closed it, and turned on the argon gas flow through the fluorinator. When the production of fluorine was fast and stable enough, John turned valves to send it through the monel tube furnace.

Professor Dickinson knocked on the door, then opened it with his key, and asked how were things going?

John's replied that there were no problems.

Later Dickinson checked on us again.

John watched the temperature and the amount of S-10 that collected in the cold trap. We then reversed the heating up process and slowly reduced both the heating current and the direct current through the pot. The fluorinator was a more difficult machine to run than the charcoal tests. The last stage was the slow one that required only one person during some of this phase; the material in the cold trap was slowly vacuum-distilled into another bulb to remove any residual HF and SF_6.

* * *

John gathered up the tools that he had borrowed for the day and took them back to the stockroom. Because of the paperwork, he would be gone for about ten minutes. I checked all the settings and took over the watch. There was little to do. I began to review what had happened that day and my thoughts branched out in many directions:

Grandpa Dial plowed and planted the garden. I had asked for and was given two rows to plant my own watermelons and corn. I started early one morning to hoe the weeds but still had not finished by mid-morning. I sat on the ground with the hoe in hand. Grandpa Dial arrived: "If a job is worth doing at all, it's worth doing well," he lectured. "Don't be lazy, my boy. Hurry up and finish hoeing your row. The longer you wait, the hotter it's going to get."

I thought about my friend, the lovely librarian at Canton High School. She, too, had rheumatic fever. In my junior year in college, I heard that she had died. I was deeply saddened for several days, both for her and for myself.

I thought about music I loved. I had left most of my records in Woodstock, because they were so heavy and fragile. I could only bring three albums, one symphony and two quartets by Beethoven, but, until recently, I had been so busy studying that I had had little time for record playing. In February the Budapest Quartet had played two Beethoven quartets in the Pasadena Playhouse, Opus 127 and 132, and I had gone, but it had cost a lot of money. Although I had repeatedly listened to these quartets on records, I had read little about them and didn't know the difference between what was great and what was supremely great. From listening to my records, I particularly liked both of these, but Opus 132 was my favorite, especially the third movement. I had listened to this movement many times. From a letter I had sent to my parents:

> It is a sin to translate Beethoven's music into words, but Beethoven himself gave words to this. He said it is a hymn of thanks to the Divinity for the recovery of a sick man. It starts softly with a slow melancholy and beautiful tune. In one form and another the tune runs through all of the movement, except for two short diversions. The tune represents the hymn of thanks. Simultaneously with the hymn the quartet as a whole describes the pain and suffering of the patient, with its strange harmonies and sonorities, loudness and softness, smoothness and agitation of the bows on the strings. A couple of times Beethoven says on the score that the suffering goes away and the patient feels better. But the pain comes back, each time more intense and nervous than before. In slow steps the music builds up to the climax. The Budapest cello player's

> black hair flew up as he yanked his bow across the strings, ripping the cat-gut of the strings and the human guts of the listeners. The beautiful tune itself shuddered at the great climax. Then the suffering slowly and irregularly tapered off, and the music ended softly. Beethoven lived through it, and wrote the quartet.

I thought to myself that day, "Actually I think the music says the patient died, the next movement begins with a forward moving nonchalant stroll: life goes on without him."

I realized that I would never have been invited to join a fraternity at Emory, except for the fact that my big brother Smith was a member of SAE, and they had to let a younger brother in. About a quarter of the members dubbed themselves the "Solid South," and they proudly and profanely expressed their reactionary and racist views. I never got along with that group. Because of my slow deliberate manner of climbing stairs or hills during the first year in college, the Solid South called me "Sneak." "Look out or he will get you. Here comes ole Sneak." I hated that name, and even in my present predicament I bitterly hated the fraternity "brothers" who called me that.

Without being called for, recollections of bad dreams I had as a child appeared.

When the next recollection began to appear, it was a recurring dream of my early puberty, whose subject was the gruesome enthymeme:

The wages of sin is death,
Sex is sin.

I slammed on the brakes to stop this line of thought. I stood up, took in a deep breath, and forced a couple of coughs, but there were no symptoms. I said to myself, "I wish it would hurry up and start, so we can get it over with. I'm getting tired of waiting around." My imagination immediately responded to this wish with a vivid day dream and in which I viewed myself as if from another person:

> When the first symptoms showed up about four p.m., all four were escorted to an army ambulance. The GI driver started up the motor and drove off to a nearby civilian hospital. A newly erected sign announced: ADMITTANCE ONLY BY PERSONS WITH SPECIAL AUTHORIZATION. Four rooms at the end of one hall were reserved. Hal was given hospital pajamas and invited to go to bed.
>
> The medical technician helped Hal during a coughing spell, and kept an oxygen mask over his face at other times. As with pneumonia in a

> child, it was not painful, except for the cough and the difficulty of breathing.
>
> When Hal coughed later on, the froth came up red with blood. Again, a vision appeared of John speaking his most terrible words, "Drowning in your own water and blood, and there is nothing anybody can do about it."
>
> Shortly thereafter, Hal died. An army administrator and Dr. Everett came into his room. The administrator asked, "What will we do about making an announcement?"
>
> The doctor answered. "Do nothing for now. It depends on what happens to the others. We need high level clearance."
>
> The administrator had another question, "Will you want an autopsy?"
>
> "Yes. This is the first human fatality from this agent, and the Corps will want to see what it has done. If possible we should find out how big a dose he took. That would be useful to us."

I suddenly cut off the daydream. With a long slender finger, I wiped a half-tear of self pity from my eye. "Useful? *Useful*, yeah! I wanted to be useful."

* * *

John would be coming back soon. I had fallen slightly behind with the S-10 work, and I quickly did the last remaining steps for this synthesis. I closed various valves, made all the final entries in the notebook, and cleaned up the hood.

John returned, put the sample in a box, full of padding on the inside, closed the lid, and we took the Stoff-Stoff down to 65 Crellin.

We exchanged greetings. John bragged about what a good yield we had. Art and the professor had finished one round of tests on the first "poof" gas and had shut down that experiment for the day. I recalled from one of the hazards sheets, "The first symptom of acid poisoning of the lungs shows up as a bitter taste in the mouth when one smokes a cigarette." From the dense haze in the room it was clear that Stosick and Dickinson had been smoking heavily, and they continued to do so. Dickinson said, "We have had a long day, but we need to stick around until six at least. Have a seat, boys. If I had some beer I would offer it to you."

Then uncharacteristically, he began to talk about himself. He was born in Brewer, Maine on June 10, 1894, the same birthday as my father but six years later. His father was a musician. His maternal grandfather was a New England sea captain. He reviewed some of his history growing up in

Puritan New England, which had some features in common with my growing up in the Bible Belt of the South. He majored in chemical engineering at the Massachusetts Institute of Technology (MIT) and received his bachelor of science degree in 1915, but found that he was more interested in pure science than in serving industry as an engineer. He was a research assistant at MIT for two years, working with the famous chemist Professor Arthur A. Noyes. Far-sighted professors and advisors to the Throop College in Pasadena had deliberately set out to upgrade the local trade school by bringing in outstanding professors from the most prestigious universities of New England. They persuaded Professor A. A. Noyes to leave MIT to come to Pasadena, and Dickinson moved with him. Throop College was renamed the California Institute of Technology, and it soon became world famous as it took on several additional star scientists from east coast universities. In 1920 Dickinson received the first Ph.D. degree to be granted by Caltech, and he was appointed Instructor in inorganic chemistry. Dickinson was a pioneer in studying the detailed positions of atoms in crystals, using x-ray methods. Dickinson had been Linus Pauling's research director. He fondly spoke about the philosophical chemical scholar, Professor Richard Tolman, who was then on leave from Caltech and occupied a high advisory post in Washington. (Tolman had written a ponderous book entitled, *Relativity, Thermodynamics, and Cosmology*, which I had admiringly thumbed through, but with little understanding).

Roscoe Dickinson and Madeline Grace Haak married on April 7, 1917, forty-five days after my father and mother were married. Roscoe and Madeline designed and supervised the building of their home in Pasadena. With their two children, Robert and Dorothy, they camped in the California deserts, where he painted the subtle colors of the landscape. He was a cello player of near professional quality. He and friends, who often came to his home, played string quartets on a regular basis. I recalled a pleasant afternoon, and smiled to myself, but the smile appeared in public. Thinking the green young student from a hick town in Georgia had no respect for string quartets, the professor challenged me, "What are you grinning about, young fellow?"

"I saw you at the Pasadena Playhouse when the Budapest Quartet was here. You were on the front row."

"The Budapest Quartet is the best in the world." That was the professor's statement. He was not interested in interrupting his own recollections, and the subject was dropped. He continued his narrative. The fellow musicians were friends, and some of their friends became his friends. He and his wife knew many people from Hollywood. Out of these many associates, a few were members of the Communist party. When the National Defense Research Committee (NDRC) had invited Dickinson to accept the contract on charcoals, he went first to the local FBI office, showed them his invitation, told them that some of his many friends, whose names he would not reveal, were probably members of the Communist party, and that he could benefit the defense effort (it was well before Pearl Harbor) by taking responsibility for this project, but that if the FBI should object to his doing so, it was no use his trying. Having come down the aisle and confessed to the FBI, they forgave him his sins.

I silently noted that somebody else has been thinking that afternoon. Reminiscences continued, as did the smoking with an occasional set of smoke rings. No one complained of a bitter taste. No one referred to why we were sitting around waiting, nor to any aspect of the S-10 spill. As time went on, I began to feel that maybe we had a chance, but it was as an impartial observer, and no elation was felt. I wondered what John and Art had thought that afternoon.

* * *

Later, a telephone call came to Dickinson. Captain Everett wanted to talk to me in the room where we had met that morning. I walked promptly up the stairs to the conference room, knocked on the door, and was admitted by the doctor. After a brief formal exchange the doctor said, "Let me listen to your chest to compare it with what I heard this morning." After he listened, front and back, "Lungs completely clear. No change at all. You and John were closest to the leaking bulb. Since you are clear, the others must be also. Tell me, where do you think that gas went?"

With a detached sense of relief, "Right up the hood, I guess."

Everett asked jovially, "Well, Hal, were you frightened today?"

"Part of the time."

The doctor chuckled, "Part of the time, eh. Well I was frightened for the four of you all afternoon."

Spirits rising, I saw an opportunity to get one up on the doctor and said nonchalantly with rural Georgia pronunciation and word choice, "We had work to do. We didn't have no time to set around being scared."

This time the doctor laughed. "O.K. You admit, I hope, it was better being busy than sitting and waiting. Dickinson thought of that. In short order, we had gotten ready for you if you had needed it. Wait here for a couple of minutes. I'll call downstairs and tell Dickinson under what terms he can release the others."

When he came back the doctor resumed his stiff professional stance, “To a high degree of probability, you are clear, but there is some small chance you might have a late reaction. If, tonight or tomorrow, you start coughing or feel any new pain in the chest, I want you to call this number at once. Don’t call any civilian doctor or hospital. Within ten minutes, one of our group will pick you up for examination and observation. I'm glad the day ended as well as it did. You may go now.”

“Thank you, sir. Good-bye.” The “sir” slipped out, but I did not try to correct it.

* * *

Back in 65 Crellin, Art and John had already left. Professor Dickinson said to me, “Are you doing anything special tonight?”

“Not really. I’ll probably finish reading a book. I am about halfway through.”

The professor continued, “I have two tickets for a play in the Pasadena Playhouse, but Mrs. Dickinson and I have other things to do tonight. Could you use our two tickets?”

“I would be real glad to use one, but I don't have a California girl friend yet.”

“I bet if you went to the Athenaeum for dinner you could find some friend who would go.”

“I bet I could too. I’ll be glad to pay you for them,” I lied, knowing how short of money I was after paying for my new music system.

“It is too late for us to turn them in for refund. They are yours free.”

With proper expressions of “thank you” and exchanges of “good night,” I left 65 Crellin and checked in my key at the sentry desk.

I went to my room at the dormitory, my roommate Fred was not there, and I wrote a standard nonconsequential weekly letter to my mother. I looked at the clock, and there was still time for dinner. Dressed in my coat and tie, I left the Graduate Student Dormitory, noting the clean air, the orange glow from the slanting late afternoon sunlight, and the steady sea breeze from the west. I fingered the two tickets in my left pocket and walked toward the Athenaeum with no thought on my mind except to hope that someone would go with me to the Pasadena Playhouse.

CHAPTER 2

EGBERT

Army representatives asked us to make enough S-10 to replace what had been lost, that is, enough to fill a small bomb. Within a month, we did it. Professor Dickinson selected a Pyrex bulb, and added attachments for conveniently filling and emptying the bulb, then thoroughly annealed it. Properly annealed Pyrex glass easily stands being immersed in liquid air. He and Art Stosick transferred the daily batches of S-10 into this bulb. When we had enough for the Army to test, Professor Dickinson packed the bulb in an insulated wooden box made to his specifications by the carpenter shop. We were allowed to follow him to his automobile, but he would not let anyone go with him to deliver the material.

When I saw Professor Dickinson's car, I exclaimed to John, "Look at that car! Have you ever seen such a beautiful car?" It was 1938 Buick convertible coupe with a rumble seat, black canvas cover, solid bright red color, white sidewall tires, and a classic radiator grill. Dickinson put the box with the S-10 in the rumble seat and closed the lid. As he drove off, I said to myself, "I hope he'll let me ride in it sometimes."

It was another month before the army got around to testing the S_2F_{10}. The explosion of the bomb took place in an enclosed barn. We heard that rats and goats had died as expected, but the explosion produced a strong sulfurous odor. Although S-10 was highly toxic, it was difficult and expensive to make; the odor produced in exploding the bomb wiped out its advantage of being odorless. Interest in S-10 ended after this test.

* * *

Professor Dickinson wrote all the reports, usually classified as SECRET. After I obtained SECRET security clearance, he allowed me to read some of the reports, and the professor explained to all of us the science and chemistry of what went on when a poison gas flowed through a gas mask. The gas might remain unaltered and be adsorbed by cracks in the charcoal particles, or it might undergo chemical reaction with water, added soda lime or added copper oxide in the charcoal. It was not correct to say that a charcoal did or did not stop a war gas. After a heavy enough exposure to any gas, any charcoal would eventually become saturated, and the gas would begin to leak through. We measured how long it took for a poison to break through, how much was retained by the charcoal at the "break time," and how the fraction of gas in the exit stream increased in time after the break started.

Before and after I joined Dickinson's project, the group had tested a long list of poisonous gases. Many World War I gases, including phosgene and mustard, contained the element chlorine, Cl. We tested many that contained chlorine, and for a time our specialty was to test gases that included fluorine, F. Consider the following three gases: phosgene, Cl_2CO; chlorofluoro phosgene, ClFCO; and difluoro phosgene, F_2CO. Of these three, chlorofluoro phosgene is the most toxic. However, we found that most fluorides react with water in the charcoals and are then strongly held by soda lime or copper oxide. In terms of amounts of gases likely to be encountered, our gas masks provided long term protection (W.A. Noyes, Jr., page 301).*

CYANIDES

We received another family of compounds, and the covering letter asked us to give them high priority in our laboratory work. These were hydrogen cyanide (HCN), cyanogen chloride (ClCN), cyanogen bromide (BrCN), and cyanogen (CNCN). Hydrogen cyanide was well known to be a fast killer, and its lethal feature is the cyanide group (CN) that it contains. Cyanogen consists of two cyanide groups, and its action is similar to that of hydrogen cyanide. At medium concentrations, the chlorine part of cyanogen cholride damages lungs and causes death within about twenty-four hours in a manner similar to phosgene; and at high concentrations the cyanide part of the molecule is a quick killer. Cyanogen bromide, which we tested, was presumably similar to cyanogen chloride, but I have no record of its toxicity.

---

*Ref. also in Preface. *Chemistry. A History of the Chemical Components of the National Defense Research Committee.* Edited by W.A. Noyes, Jr. An Atlantic Monthly Press Book, Little, Brown and Company, Boston, 1948, 524 pages.

General ref. in this chapter. Marschall Sittig, *Handbook of Toxic and Hazardous Chemicals*, Noyes Publications, Park Ridge, New Jersey, 1981.

When these poisons arrived, John and I discussed how we could use our method to measure the amount of the gases. We concluded that our method would not work for these four compounds, and Art Stosick agreed with us. We determined the amount of gas breaking through the charcoal by measuring the change of electrical resistance of water after the gas was dissolved in it. Pure water has high electrical resistance. Ions in water are good conductors of electricity, which we measured as a low electrical resistance. Table salt (sodium chloride, NaCl) dissolves in water to give positive sodium ions (Na^+) and negative chloride ions (Cl^-), and this solution is a good conductor of electricity. In general, salts, for example, potassium chloride (KCl), calcium chloride ($CaCl_2$), aluminum chloride ($AlCl_3$), and sodium nitrate ($NaNO_3$), dissolve in water to give positive and negative ions, and these are good conductors. Hydrogen chloride (HCl) reacts with water to form charged ions, hydronium positive ions (H_3O^+) and chloride negative ions (Cl^-), which provide excellent electrical conductivity. Hydrogen cyanide swims around in water as a three atom molecule H–C≡N, and it is a poor conductor of electricity. Cyanogen exists in solution as the molecule N≡C–C≡N, and this solution does not conduct electricity. Cyanogen chloride reacts with water to give hydrogen cyanide and hypochlorous acid (HOCl), which forms only a few ions in solution and is then a weak conductor. Even if these compounds were passed through a red-hot tube, the decomposition products would not give good signals in our apparatus. Professor Dickinson solved the problem as soon as we mentioned it to him. He said that instead of using distilled water as our primary reagent, we should use a dilute solution of mercuric chloride ($HgCl_2$) dissolved in distilled water. If the professor had asked me, "Would a solution of mercuric chloride in water be a good conductor or a poor conductor," I would have answered that it would be a good conductor, like calcium chloride, but that is wrong. As a major exception among salts, mercuric chloride dissolves in water yet retains its molecular form, Cl–Hg–Cl, and its solution is a poor conductor of electricity. The professor's recommendation was even more subtle than it seemed to be at first. He explained to us that mercury powerfully attracted the cyanide group and would react in solution to form ions and be a good conductor. A simple example is the reaction of hydrogen cyanide with mercuric chloride in water. The mercuric chloride molecule snatches the cyanogen from first one and then another hydrogen cyanide molecule, releasing hydrogen chloride in the process. As noted above, hydrogen chloride reacts with water to produce ions in solution. When all of these reactions are added together, we see what happens in our system;

net: $Cl{-}Hg{-}Cl + 2\ H{-}C{\equiv}N + 2\ H{-}O{-}H \rightarrow N{\equiv}C{-}Hg{-}C{\equiv}N + 2\ H_3{}^{+}O + 2\ Cl^{-}$

Mercuric chloride and hydrogen cyanide and water react to form mercuric cyanide and hydrochloric acid ions.

All of the reactants are poor conductors of electricity, but the positive and negative ions produced by the reactions are good conductors. With this ingenious trick, we were able to use our method on all of these poisons.

WEEKEND RECREATION FOR CALTECH GRADUATE STUDENTS AT SOUTH LAGUNA BEACH

Carl Niemann was professor of organic chemistry at Caltech, and later was active in the NDRC chemical warfare research. Several of his graduate students in 1941 and 1942 rented an old house at South Laguna Beach for thirty five dollars per month. James F. (Jim) Mead, a third year graduate student in Professor Niemann's laboratory, was a beach volley ball enthusiast, and he organized the weekend expeditions to the beach. The house was furnished, and had double bunk beds for six people and floor space for additional sleeping bags. Typically a car load of graduate students in Niemann's group spent Saturdays and Sundays there.

The group developed a tradition of shared chores. Half of them planned the meals, shopped for food, and did the cooking, and the other half washed the dishes and cleaned the house. On the next weekend, they might reverse these roles. There was some effort by the cooks to prepare a "gourmet" dinner for Saturday nights, and there was teasing about the outcome of these efforts.

Figure 2.1. shows four Caltech graduate students in the fall of 1941 racing on Laguna Beach: John Hays, James F. Mead, Richard Newton Lewis, and Andy Benson. Andy Benson lived in the old graduate student dormitory for the first two quarters of the school year, as did I.

In the spring of 1942, several of Niemann's graduate students completed their Ph.D. degrees, and others became occupied with other things. Andy Benson finished Caltech in the Spring of 1942, was married in May 1942, and accepted an appointment as instructor at Berkeley. (He played an important role in the events of Chapter 3). To fill the weekend cottage and to pay the rent, other graduate students in chemistry joined the few who remained of the original group. I was invited to go to the beach one weekend in June. The cost for food, lodging, and transportation was two dollars per person. The group consisted of two loaded cars. Half of the people, including me, slept on the floor. With great interest I explored the beach for a couple of miles; I was especially

Figure 2.1. Caltech graduate students racing on Laguna Beach, Fall 1941. Photograph supplied by Andy Benson.

interested in the large rock formations and the tide pools. The beach was different from the long sandy beaches of Georgia and Florida.

Linus Pauling had persuaded Professor Lazlo Zechmeister, a great organic chemist from Hungary, to accept a permanent position at Caltech, and he brought with him his essential assistant, Dr. Andy Polgar, who was only a few years older than the graduate students. Andy Polgar was one of us at the beach that weekend. Andy had been to this beach party several times before. He had admitted that he was experienced at cooking, and had agreed to cook a Hungarian pot roast of beef for the group. During the preceding week, he went to a store and bought a fine piece of beef. Paprika was heavily used in Hungarian cooking, and it was a major source of vitamin C in their diet. Andy's command of spoken English was not strong. Andy selected and bought a big bottle of what he thought was paprika, but it was ground red hot pepper. At the beach house on Saturday night, Andy roasted the beef and heavily coated it with the red powder. Andy proudly sliced it and served the group with his gourmet dish, and quickly everyone was gasping and coughing at the hot red pepper.

Figure 2.2. Professor Don Yost, Caltech.
Photograph supplied courtesy of California Institute of Technology Archives.

The meat was inedible.

The trip back to Pasadena on Sunday afternoon was memorable for the traffic jam. We slowly drove bumper to bumper along the two-lane highway. Occasionally the road had a right-angle turn, and as we approached within sight of such a turn, we could see a continual string of cars creeping along the road.

PROFESSOR DON YOST

Professor Don Yost's last graduate student, Terry Cole, wrote in the Caltech magazine: "Tribute by Terry Cole," ENGINEERING AND SCIENCE, Vol. 41, No. 1 28–29 (1977).

> ...Don's pioneer youth had a profound influence on his character and unique approach to science. He was born in the village of Tedrow in northwestern Ohio. By 1899 economic conditions forced his father to give up farming there and move, first to the lumbering camps of northern

Wisconsin, and finally, in 1902, to a ranch in the Boise Basin of southwestern Idaho.

Don's often-interrupted education continued at a frontier school near the ranch. Its enrollment consisted of about ten children and a half dozen wintering cowboys. He once remarked that the lessons were far from memorable, but the exhibitions of fancy horsemanship by the cowboys at noon recess were always exciting. During high school Don acquired his enduring fascination with mathematics and languages so familiar to later generations of his students...

In the summer of 1914, his acceptance in hand and the $10 out-of-state tuition paid, Don arrived, via rail and steamship, in San Francisco to begin his college education at UC Berkeley. His freshman year was decisive; by the summer recess he had found his calling through the inspired teaching of his chemistry professor, Joel Hildebrand and a young lab instructor Richard Tolman.

During his second year Don met, and in the following year married, Susan Marguerite Sims, later affectionately known to his students as Mamacita. A month after their marriage, the United States entered World War I, and Don enlisted in the Navy, where he served for three years. He graduated from Berkeley in 1923...

He applied for graduate work with Arthur A. Noyes at the fledgling Institute. His career at Caltech was brilliant and wide ranging. Upon receiving his Ph.D. (magnum cum laude) in 1926, he was appointed instructor in inorganic chemistry and began the application of the most modern physicochemical techniques to the elucidation of the chemistry of the rarer elements... His work on the volatile fluorides brought him international recognition...

During the 1930s, Don published over 50 papers: contributions to chemical kinetics, gas equilibria, the chemical effects of X-rays, electrochemistry, the chemistry of the platinum metals, low temperature thermodynamics, and rare-earth chemistry. His achievements during this time are the more outstanding when viewed in the historical context. In those years there were no high-technology instrument manufacturers; any apparatus more complex than a galvanometer or simple glassware had to be built or improvised as the research went along.

Soon after the formation of the National Defense Research Committee, Don was sought out to direct war research He was appointed Section Chairman under the OSRD (Office of Scientific Research and Development), directing research teams at Caltech, Northwestern, and Los Alamos (and Dugway Proving Ground). His achievements in this capacity were to bring him the Presidential Certificate of Merit.

Change of the nature of our research

In the summer of 1942, the National Defense Research Committee (NDRC) changed completely the nature of our research. In July 1942, a group of university scientists who were working on chemical warfare problems met in Evanston, Illinois; this group later became Divisions 9 and 10 of the National Defense Research Committee (NDRC). By that time many potential chemical war gases had been identified, synthesized by chemists, tested on animals for toxicity, and tested with respect to how well gas masks stopped them. The group discussed what else should be done.

Professor Don Yost was short, slender, and spoke with a calm clear voice in his class in advanced inorganic chemistry in the fall of 1941. At this meeting, Yost spoke forcefully to the effect that we had overcome the emergency we faced in 1940 of our having poor quality charcoals and inadequate gas masks, that we had solved some of our most urgent defensive problems, and that we should understand how to carry out offensive gas warfare. Professor Yost said we should be ready to start and carry out chemical warfare if that was to our advantage. He added that we did not know how gases would perform as determined by wind, temperature, time of day, cloudiness, and different locations. Professor Wendell Latimer of the University of California supported Yost's position, and it became the policy of NDRC Division 10, which was formed soon after the meeting at Evanston.

The U.S. Army had not carried out field research of this sort since World War I. Such research had been done at Suffield in England, but only over a narrow range of terrain. Our army had recently set up the Dugway Proving Ground at Tooele, Utah, but it was designed to prove the performance of chemical weapons, not to understand what happened to the gas in the field. Yost was appointed Chairman of this section of NDRC, and he initially directed this line of research at Dugway, University of Illinois, Northwestern University, the University of California, and Caltech. (Noyes, pages 318–319).

Professor Dickinson approved our acquiring the ability to carry out offensive chemical warfare as the best defense against the enemy initiating such warfare ("We are guarding a bridge," he had said, and he hoped it would not be attacked).

OUR NEW LINE OF WORK

The goal of our new line of work was quantitatively to understand the effect of terrain and weather on the persistence and danger of war gases. Our first job was to learn something about meteorology and to design and build portable instruments to measure air temperature and wind speed out-of-doors from ground to five meters height. Such ground level meteorology was called "micro-meteorology." We were to obtain pertinent micro-meteorological measurements near potentially vulnerable cities from the Mexican border to San Luis Obispo to see if there were any regions especially vulnerable to gas attack. Professor Wendell Latimer and his group at Berkeley were to make similar measurements from San Luis Obispo to the Oregon border.

New cars were not available for non-war work, but Dickinson was able to buy a new 1942 Buick station wagon for our war project. It had real leather seats, and real wooden panels on the outside. It was an impressive automobile. It had two fold-up seats between the front and regular back seat, so the car could carry eight people. The back seat could be folded down to give a large storage area. We loaded additional equipment in the storage rack on the top of the car.

SOME MICRO-METEOROLOGICAL CONCEPTS

The normal situation in the lower atmosphere is for both air temperature and air pressure to decrease with increasing elevation. During daytime, the sun heats the earth's surface, air in contact with the surface is then heated, becomes lighter than the air above, and will tend to rise as invisible bubbles and swirls, but which are sometimes seen as "dust devils." Under these conditions a cloud of gas at ground level is carried off by the wind and rapidly mixed up and down. At night the earth continues to radiate heat to outer space, and thereby surfaces become cooler. Air in contact with the chilled surface becomes colder and heavier than higher air and acts almost as a separate fluid. Surface-cooled air forms stagnant pools on flat surfaces and flows down-slope from hills and mountains, a "katabatic wind." This increase in air temperature with elevation is called a temperature inversion, which confers great stability to air against vertical mixing. (The technical definition of a temperature inversion is slightly different from this simple statement, but the difference is not important in this discussion). A cloud of gas in air with a temperature inversion may be transported by wind for large distances horizontally, but it mixes slowly with air above it. A given amount of a war gas can be hundreds of times more lethal at night than it is by day.

We discovered that the Los Angeles basin was especially vulnerable to gas attack. The average wind speed in the Los Angeles basin was low. During much of the year there was a pronounced breeze from the ocean to the interior during the day. This large-scale prevailing wind, which at its lower elevation had been chilled by contact with the cold ocean below, moved air into the Los Angeles basin. The cold ocean air slips like a wedge below the warmer land air, thereby creating a large scale temperature inversion, which is a few hundred to a few thousand feet deep over the Los Angeles basin. Although the meteorology of the Los Angeles Basin is complicated, its vulnerability to gas attack or to air pollution can be basically understood in terms of weak prevailing winds, frequent temperature inversions produced by advection of cold shallow layers of ocean-chilled air and by radiative surface cooling at night.

When winds move upslope, the air expands and cools ten degrees Celsius per kilometer or five and a half degrees Fahrenheit per thousand feet. When moist air is lifted it cools at the same rate as dry air until the temperature falls to the dew point, at which point water vapor begins to condense to form clouds. When winds move downslope, the air temperature increases ten degrees Celcius per kilometer or five and a half degrees Fahrenheit per thousand feet, and lateral expansion of this heated air increases the wind velocity. In the Los Angeles area, these hot dry winds are called "Santa Anna winds." Such winds have other local names over many regions of the earth.

In 1941, there were not enough meteorologists in the country to provide service for the Air Force, and thus only a few real meteorologists could be spared for work with the Chemical Warfare Service. The Air Force set up special courses for enlisted men who had enough knowledge of physics and mathematics to start learning meteorology so there was no loss of time reviewing background material. Students who passed the course were eligible to be Second Lieutenants in the Air Force. Caltech taught such a course to several successive groups during the war. I read much of the textbook, the instructor let me have copies of his examinations, and he gave me some weather maps. The instructor of the course gave a trick question in an examination at the Air Force meteorology school at Caltech. He passed out a weather map of and near Greenland. There were no storms or storm systems on the map. Large scale pressure gradients were weak, and there were no clouds. The student was asked to predict the wind direction and speed for the next twelve hours. I took this examination, and like all students I talked to, I answered "light winds, variable direction." That answer was wrong. The actual situation was a gale force wind off the body of Greenland. This gale was caused by katabatic winds. Outgoing heat radiation through the clear air dropped the temperature of the ice surfaces, additionally chilling the cold surface air. This cold air built up deep stagnant layers over relatively flat surfaces and flowed like rivers down the mountains, down the valleys, merged with other air torrents like a flood, and swept over the weather station with almost hurricane force. Every student got the wrong answer; this problem was designed to maintain restraint and humility in young meteorologists and to teach them about katabatic winds.

British micrometeorologists had made extensive measurements of the ratio of wind speed at the two meter height to that at the one meter height, and they named this ratio "R-value." They had shown that wind speed and R-values were significant variables to use in characterizing the dispersion of war gases over smooth grass fields.

MICROMETEOROLOGICAL INSTRUMENTS

Measuring air temperature out of doors is not a trivial task. If a bare thermometer is placed in the air, it will absorb sunlight and read too high a temperature. If the thermometer is shaded from the sun, it will absorb infrared rays from the ground and again give a spurious temperature, too high or too low. If the thermometer is shaded above and below, the shades may absorb sunlight or ground heat and change the local air temperature. To obtain true air temperature, we mounted a small copper-constantin thermocouple within two thin-metal, concentric tubes and rapidly aspirated air through the tubes and across the thermocouple. We built a vertical mast five meters high made of an aluminum pipe of four-inch (10 cm) diameter in three detachable sections, and we attached our aspirated temperature-measuring instruments to the mast at one-third, one, two, and four meter heights. We used a direct current motor powered by two storage batteries to drive a large vacuum-cleaner and hooked it to the base of our vertical tube mast. These temperature measuring masts were later manufactured by the Wheelco Instrument Company and used by other groups.

As a reference for the thermocouples, we used mercury-in-glass thermometers, graduated to tenths of a degree, calibrated against a National Bureau of Standards thermometer. One day I accidentally broke one of our two thermometers. This destruction of government property greatly upset Professor Dickinson. "How can I carry out a Defense Project if the equipment is going to be carelessly broken?" After he calmed down, he gave advice, which was: "When you are spending Government money, always consider how your actions will sound in a Congressional investigation five years from now."

CONSTRUCTION OF NEW INSTRUMENTS

To measure wind speed, we bought a set of commercial cup anemometers, but they gave results in terms of numbers on a dial. It was a nuisance to use a stop watch to start and stop the anemometer readings, and then to write the number in a notebook. We designed and had the Caltech machine shop build an electrical recording wind speed meter. We mounted three light aluminum anemometer cups on a vertical axis that spun on a jewel mount, and at every complete round, a small pointed needle contacted a mounted drop of pure liquid metallic mercury to give a brief electrical pulse. Art Stosick hooked together a set of vacuum tubes that counted the pulses to give a continuous recording of the wind speed. The Lane-Wells Company manufactured thirty-six of these instruments for other groups.

The Caltech machine shop built a few British "gustiness meters," which was a wind vane free to move up and down as well as left-to-right. It had a light-weight ink pen on its downwind side, which put ink marks on a sheet of curved paper. We let it operate for two minutes, and the pattern of scribbles indicated the degree of horizontal and vertical turbulence.

We packed red phosphorus and another powder in cardboard cylinders about three centimeters in diameter and fifteen centimeters tall; when ignited by a thin magnesium ribbon as a fuse, one of them produced a copious amount of smoke for a few minutes. Heat from the hot burning mixture lifted the smoke up off the ground, and as the wind moved the smoke along, it could be seen to spread in vertical and horizontal directions. It spread rapidly both up and down during the day, when there was an unstable lapse rate. During a temperature inversion, the smoke would boil up a few feet above the ground and then slowly spread gauze-like in broad flat sheets with little vertical mixing. We photographed these patterns of motion. Thus we had four different ways to characterize air turbulence near the ground: temperature and wind speed vertical profiles, the gustiness meter, and smoke patterns.

To develop our novel micro-meteorological instruments, we used a large cow-pasture east of Pasadena, which gently sloped up toward the San Gabriel mountains. We had a favorite demonstration of katabatic wind. We had a pole with eight wind vanes spaced about every half meter apart. In the afternoon, the sea breeze caused all the wind vanes to point toward the Pacific Ocean. As sunset approached, first the lowest wind vane on our pole turned and pointed toward the mountain as a shallow current of cold air flowed downhill. Somewhat later the second vane turned, then the third, fourth, fifth, sixth, seventh, and finally the eighth turned as the katabatic wind built up to be a deep gentle flow of cold air. Before sunset, we could almost tell the time of day by how many wind vanes pointed uphill.

We built a complete, mobile meteorological station that could be dismantled and stored in the back of and on top of the new 1942 Buick station wagon. We used these instruments to measure local conditions at the beach, in a desert, on a site up the side of the mountain, and at various spots in the city.

A HIDEOUS QUESTION

A chemical warfare Colonel visited our group, interviewed Professor Dickinson in his office, and informally interviewed the rest of us in the laboratory and in the cow-pasture. In the afternoon of the second day of his visit, he and I sat talking on the marble steps of the east entrance of Gates Chemical Laboratory (Figure 1.3.). I forget what we talked about until he asked me a startling question: "I don't want to go through channels. I want you to give me a rough answer to an important question. How much phosgene would it take to kill half the people in Los Angeles?"

I was struck silent by this hideous question. The Colonel waited. I realized that he was not going to gas Los Angeles, and he needed the information to plan our defense. I finally asked, "What is the LD50 of phosgene?" LD50 is shorthand for lethal dose to kill fifty percent of the exposed population.

He said, "550 ppm-min," (550 parts per million of phosgene in the air multiplied by minutes of exposure). I wrote down the value and replied I would think about it that night and talk to him the next morning.

My first thoughts that night focused on the overwhelming difficulty of the problem. Houses were built on a variety of terrain: hilltops, slopes, level areas, near the ocean, and inland beyond the first row of hills. Already from our measurements, we knew that katabatic winds at night flowed in different directions and speeds and attained different depths in the Los Angeles basin. If gas were dropped as bombs, there would be great differences in exposure close to and far from the bombs. Would there be a surprise attack or would people be warned by radio to go indoors, shut all doors and windows, and stuff all cracks with wet rags? I recalled one line of poetry by Louis Carroll, "The farther off from England, the nearer is to France." The complex winds inside the region would replace bad air in one place with bad air from another place, but these details were not significant to the problem. The question concerned what might happen over a large area. My answer could not include an average over the myriad of complex details, but had to be a general average reached by thinking of the problem as a whole. Probably the worst situation for a poison gas attack on the Los Angeles Basin would be during a cold cloud-free night with a weak general wind. There would be a strong temperature inversion caused by thermal radiation from the land and cold airflow down the hills, which could expose a large population for a long period. How many hours? I assumed two hours. "Dose" is defined as the product of concentration, C, of gas multiplied by the duration of exposure, t. Dividing the colonel's LD50 by two hours gave the concentration required (literally, this is the mole fraction, not the concentration). The mass of phosgene required would be proportional to the depth of the cloud. Not knowing what a real case might be, I assumed a range of depths of twenty, fifty, and one hundred meters. With my slide rule I came up with some numbers. I do not remember the assumptions and numbers I used, and so I made the calculation again in 2001 to illustrate the sort of thing I remember doing.[#]

The next morning I explained to the Colonel the logic and emphasized the uncertainty of what I had calculated. I showed him three numbers for mass of phosgene required, depending on the three assumed depths of gas: 1000 tons, 2000 tons, and 5000 tons. During daytime or with a moderate wind at night would require much more. The Colonel said that my numbers were the same sort he had arrived at, but he wanted "one of you bright guys to verify my guesses." He said that some people had been concerned that a fleet of submarines could slip up and release their tanks of gas from the nearby ocean, but they could not possibly put out that much phosgene, even for my case that indicated the smallest amount.

[#]Consider one case. Assume area of Los Angeles is that of a square 50 kilometers on edge, depth of gas cloud is 20 meters (worst case), exposure time is two hours, temperature is 293 K, LD_{50} is 550 parts per million × minutes (where I assume parts per million is a measure of mole fraction, not of mass ratio), and molar mass of phosgene is 98.9 grams. The calculation in 1942 was done with cgs units. The calculations are:

Volume of air = $(50 \times 10^5)^2\ \text{cm}^2 \times 2000\ \text{cm} = 5.0 \times 10^{16}\ \text{cm}^3$
Moles of air = $5.0 \times 10^{16}\ \text{cm}^3/25{,}000\ \text{cm}^3$ per mole = 2.0×10^{12} moles. (Ideal gas law).
ppm of phosgene = 550 ppm × minutes/120 minutes = 4.58 ppm
Moles of phosgene = moles of air × $\{4.58\ \text{ppm} \times 10^{-6}\} = 9.2 \times 10^6$ moles
Mass phosgene = 9.2×10^6 moles of phosgene × 98.9 g $\text{mole}^{-1} = 9.0 \times 10^8\ \text{g} = 9.0 \times 10^5\ \text{kg}$
= 900 metric tons ≈ a thousand American tons

Reference for 550 ppm × minutes: *Phosgene Toxicity,* A report of the Major Hazards Assessment Panel Toxicity Group, INSTITUTION OF CHEMICAL ENGINEERS, Davis Building, 165–171 Railway Terrace, Rugby, Warwickshire CV21 3HQ, UK, 1993 Chapter 2, p. 11.

That case would require a fleet of aircraft twenty times larger than the attack on Pearl Harbor to kill half the people in Los Angeles with phosgene. "Such a major effort," he said, "would have a bigger pay-off if explosives were used."

MY BEAUTIFUL NEW CAR

In October 1941, I had gone to the bank at the intersection of Lake Street and Colorado Avenue, to open a bank account and deposit money my parents had given to me. The teller referred me to a bank officer, who asked me several questions and discovered that I was poor graduate student at Caltech. Apparently he had faith in the future prosperity of Caltech students, and he cordially accepted my small amount of money and encouraged me to do business with his bank. At the end of the 1941–42 school year, I had to move out of the graduate student dormitory, since I was no longer a graduate student. I obtained a room in Mrs. Matt Thompson's boarding house at 1135 Constance Street, a short two blocks from the Gates laboratory. From my salary of $125 per month, I was able to save about $75 each month. During the early summer of 1942, I did get to ride in Professor Dickinson's gorgeous Buick convertible coupe, in the front seat with the top retracted on one occasion and in the rumble seat on another. In the Fall of 1942, I wanted to buy a car of my own, and I had $380 in my bank account. I studied the want ads in the newspaper, looked at a couple of cars in my price range, but turned them down.

One day a short entry in the want ads offered for sale a 1937 Buick for $450, and even though I did not have quite that much money, I reached the address by streetcar. The owner was about my age, he had just been drafted, and his father was selling the car for him. He showed it to me parked on the street. It was a 1937 Buick convertible sedan with white sidewall tires and a hood pattern the same as that of Professor Dickinson's 1938 Buick. It was painted a dull green color and the roof was an undistinguished tan, but a lack of surface beauty did not matter. I drove the car around the block and strongly wanted that car. I wrote out a check for $50 as a deposit, and he agreed to give me four days to bring him the rest of the money. The next day I went to my bank, stood in line to reach a teller's window, and said I had an account there and wanted to borrow $150. The teller looked up my account, and there must have been some note there left by the officer who opened the account. The teller made a telephone call, and then told me that in fifteen minutes I would have an appointment with the chief bank officer who handled making loans. I reported to him on time. The chief bank officer sat in a leather chair behind a large polished hardwood desk. I said I wanted to take out a loan for $150. He was cordial, serious, and strictly did his business. (Later I concluded he was taking time out to have some fun). He asked about my job and what was my salary. He noted my bank account and that I had been saving about $75 a month, which he approved of. I explained my chance to get an automobile of my own and that I could pay back the loan in two months. He said, "If we give you this loan, what will you put up for collateral."

"Please, sir, what is collateral?"

"Well, some of our clients in getting a loan, deposit stock with us or a deed to some property they own."

I felt I had no chance to get the car and that I would lose the $50 deposit, but I managed to say, "Sir, I don't have any collateral."

His serious reply was, "Well we could take the deed to your car as collateral. If you don't pay up, we get the car. But I advise you to give yourself three months to pay off the loan. You don't want to risk a cash-flow problem."

We filled out a formal document; probably the sort he would use for a large business loan. He calculated the interest on the loan, added it to the total, and said it would show up the next day in my bank account. I expressed gratitude to him. He grunted, we shook hands, he went back to work, and I walked out happy.

I posed for a picture along with my car (Figure 2.3.).

For six dollars a month, I rented a one-car garage at the rear of a narrow vacant lot, which was about a half block away from the boarding house where I lived. During the war, gasoline was rationed, any car owner could get an "A" sticker and coupons for a few gallons of gasoline per month, and drivers who needed a car in support of their business or the war effort had "B" or "C" or other stickers and received more coupons for gasoline. Since I used the Buick only for recreation, I received the minimum number of coupons. During the next three years, I was away from Pasadena for extended intervals of time. Before I would go away, I left the car with a good friend for its care and exercise. When I got back, my friend had gotten engaged to be married, and this happened to three friends. They were:

Bill Lipscomb, later, Professor of Chemistry at Harvard and Nobel Prize winner.

Arthur Pardee, later, Professor of Biological Chemistry and Molecular Pharmacology at Harvard, and did outstanding research at Dana-Farber Cancer Institute.

Figure 2.3. Hal at Caltech Athenaeum with his 1937 Buick convertible sedan, 1942. Photograph given to HSJ by unknown photographer.

Bob Mills, later, distinguished research scientist at Los Alamos National Laboratory.

DUGWAY PROVING GROUND, TOOELE, UTAH

Dugway cordially welcomed NDRC people. They positively wanted competent scientists to visit their site and to understand their needs. Several of us on Dickinson's and Yost's projects at Caltech took turns in visiting there. Professor Dickinson and I went there by train in mid December 1942, Dickinson left shortly, and I remained at Dugway until early January. Dugway Proving Ground consisted of two hundred and fifty thousand acres (hundred thousand hectares) of largely flat salty soil in Utah. In the December clouds and haze, I could see no horizon; everything was the same dull white in all directions, including up and down. The buildings had icicles over the outside and were overheated by coal burning stoves on the inside. I was a low man in the organization and was assigned to work with scientists, both soldiers and civilians, in their chemical laboratory. I lined up for chow with the soldiers, as did the civilian laboratory workers.

One day I was taken for a jeep ride over the testing field. We passed a twisted metal object with liquid in its concave portion. The soldier driver steered clear of it and said it was the remains of an air-drop bomb and the liquid was residual war gas. It was December and quite cold. The laboratory crew told me about a big experiment Dugway had recently performed. The Army shaved the wool off the backs of a flock of sheep, and in an area remote from the headquarters built a fence to contain the sheep. Aircraft flew overhead at an unspecified elevation and sprayed a large amount of Lewisite (one of the World War I persistent gases). The sheep were examined but showed no damage. Apparently, the Lewisite evaporated before reaching the ground.

For several days I explained to the laboratory crew of soldiers and civilians how we measured gas concentrations by electrical resistance in a water solution, and they planned to set up such a system. Each Saturday morning the laboratory staff stopped their work, dismantled their apparatus, cleared everything off the laboratory benches, washed and dried the table tops, and swept the room. The commanding officer carried out his weekly inspection. He walked into the room accompanied by two junior officers, quickly glanced around the room, wiped a finger over the top of the laboratory bench, noted that his finger was clean, and marched out of the room.

Christmas at Dugway was no special affair. Some work went on during the day. We had a generous turkey dinner that night, and afterwards the soldiers and we laboratory technicians lined up and filed in to see a highly rated new Hollywood movie. The wooden benches had no back rests. There was no comic short piece as introduction. The movie, starting with a flair of pretentious music, was "King's Row." The parts I remember involved an idealistic young doctor (name forgotten), his flamboyant care-free friend (Ronald Reagan, I never forgot him), an honest sympathetic middle-age doctor (Claude Rains), his beautiful daughter (name forgotten) in love with the young doctor, a crazy woman locked in an attic, a sadistic surgeon, and the good girl (Ann Sheridan) who lived on the wrong side of the tracks. In the middle of the movie, Claude Rains murdered his daughter and committed suicide. Later the sadistic surgeon needlessly cut off both of Ronald Reagan's legs at the hip after he came in for treatment following a minor accident. This was a gruesome conclusion to a lonely Christmas.

My train on the way home was stopped for three days in a desert near Yermo, California. A heavy winter storm had hit the desert region of southeastern California, set off flash floods in the gullies, and washed out the train track. Buses finally arrived and took us home to Pasadena.

SMOKE TESTS

A few lush orange groves remained in the city of Pasadena, and vast areas of land east of Pasadena were covered with orange trees. Even in the balmy Los Angeles basin, occasional spells of frosty weather came with conditions cold enough to spoil oranges on the tree or even damage the trees. To protect the fruit and the trees, farmers installed a series of smudge pots, which consisted of an oil tank as the base and a burner mounted above the oil tank, rising about five feet above the ground. The oil burned to give a heavy dark smoke, which protected the trees from frost.

In Los Angeles there were large aircraft factories. Tall chain link fences, with a tall sloping barbwire barrier on top, were placed well away from the manufacturing facilities to protect the aircraft factories from intruders. Armed soldiers patrolled and protected the fences. The Army had requisitioned farmers' smudge pots, and placed them over wide areas near and far from the factories. One duty of the soldiers was to rush out and ignite the smoke pots if there should be an approach of enemy aircraft. Occasionally the soldiers were called out to light the pots as test for their readiness and for observers to note how well such smoke clouds would protect the aircraft factories.

Through some negotiations, the army invited Dickinson to set up our meteorological instruments in a field outside the fence that surrounded an aircraft factory on a day when they planned a surprise activation of the smoke pots. In April 1943, we four drove up in our new Buick station wagon about a hundred yards from the factory fence, got out and erected our five-meter-high aluminum tubing that we used for our temperature measurements, set up a another post with two anemometers and installed our other instruments. We started to make our measurements and planned to continue until the smoke pots were lighted and the sky filled with smoke, which was scheduled to start in about half an hour. Our attentions were so concentrated on our job that we did not notice developments occurring around us. Suddenly, we found ourselves surrounded on three sides by armed soldiers with fixed bayonets. The officer in charge with three enlisted men walked up to our site, and Professor Dickinson walked out to meet them. The officer barked out something. Dickinson said something to the officer, but in order not to let the soldiers know about the surprise order to ignite the smoke pots his remark was unsatisfactory. The officer barked again, and the soldiers, now all around us, slowly approached. Dickinson called the officer aside and with a low voice rapidly spoke to him and showed him a piece of paper. The officer called to the troops to halt and back up to about fifty yards away, where they remained and watched us for the rest of the time we were there. The officer put in a call on his "walky-talky." This was a new device, a portable wireless telephone, carried by a strap across the shoulder in a substantial leather case. We continued taking measurements, soon the order came to ignite the smokepots, and the sky filled with dark oily smoke. Eventually, there was an all-clear signal, and we left.

BOB MILLS

In May 1943, Bob Mills joined Dickinson's project. Bob was born on June 6, 1922 in Canton, Ohio. He received a scholarship at Washington and Jefferson College in Pennsylvania, where he made excellent grades and was a star baseball player throughout his college years. Bob was skilled in many activities, jovial, and pleasant to talk with. He readily showed a grinning face and smiling eyes, and he was best friend to almost every person who knew him well (Figure 2.4). He soon became a vital member of our team.

John Otvos and Bob Mills sponsored, encouraged, and directed the Caltech chemists baseball team. They practiced in Tournament Park next to the Caltech campus and played serious games in Brookside Park, which is near the Rose Bowl, against several other Pasadena groups of the Owl League. There was serious competition among the groups and heavily fought playoff games toward the end of the season. The chemists on the team became a close knit fraternity and many of them remained good friends for life. (Needless to say, I could not play baseball and was not a member of the fraternity). Rene Scott, later Rene Mills, attended all the games and recorded every play of every inning. Rene typed up these records for about three years and bound them in one large book. She kept statistics on each player, such as times at bat, hits, runs, and errors. Her record of the game discussed here took seven singled spaced typed pages. I quote here a sample of her account, which is the description of one half of one inning:

> As the Chemists came in for the last of the ninth [the score was tied] it was apparent that the game could not last much longer. Both teams were scoring, and it was only a matter of time until somebody broke through. While Senear warmed up Mills to take over the pitching duties if a next inning should be necessary, Lanni started things off by grounding sharply to second. Than Lanky Dick (Lemmon) strode to the plate, determination written on his face. It is said that he asked "Will one run win?" but this is probably apocryphal. He watched one pitch go by, and then came the payoff, the most exciting and thrilling event in the Chemists

Bob Mills

Figure 2.4. Photograph 1942 supplied courtesy of Rene Mills.

> long history. On a shoulder high pitch over the center of the plate Dick uncoiled his long, lean body, and smote the ball a terrific blow. At first it looked like a long fly, as Tarzan in left field drifted over for it. But it suddenly became very apparent that the ball was going like a shot out of a gun and Tarz could not reach it. In fact it went past his outstretched glove with feet to spare. Then came a half minute that seemed hours, as everybody watched two runners, Tarzan disappearing into the outer darkness in left field and Lemmon rounding the bases. He took eons going from first to second. Coming into third all eyes focused on Otvos to see if he would hold him up, but Dick was waved on in, crossed the plate standing up, and the game was over. Actually the relay from the outfield never did get in, being lost behind third base someplace, but it never could have been even close as Dick was halfway home before the ball even put in an appearance.

The Pasadena newspaper gave an account of the game:

"Chemists Top Eagles, 5–4, in Owl Thriller

"In the most exciting game of the Owl League playoffs, Caltech's underdog Chemists dumped the strong Fraternal Order of Eagles nine, 5 to 4, in a thrilling contest at Brookside last night.

"The Chemists came from behind twice. They counted three runs in the seventh and won the game in the last of the ninth when Pitcher Lemmon wacked a long homer over the left fielders head to win the game.

"Cornith and Wright starred for the F.O.E. team while the entire Chemist lineup played heads-up ball."

ART PARDEE

Dr. Malcolm Dole was professor of physical chemistry at Northwestern University. He supervised an NDRC group at Dugway in early 1943. In his written evaluation of his NDRC crew, he said concerning Arthur Pardee:

> Arthur Pardee was another valuable member of the group being strong physically and brilliant mentally. He designed and directed the construction of the Japanese type bunkers and dugouts for the Florida Field Trials. On many other special problems his assistance proved to be excellent.

Arthur Pardee graduated from Berkeley with a bachelors degree in chemistry, came to Caltech as a graduate student, and described some of his experiences at Dugway:

> In 1942 I entered CALTECH, where I did research on immunochemistry in the Pauling group. Earned my MS by 1943.
>
> I joined the NDRC chemical warfare project in the Summer of 1943, and went to Dugway Proving Ground. My boss there was Malcolm Dole from Northwestern U. Others there I remember are Paul Saxer, Norman Larsen, Howard Pinkard, Nelson Neese, Stu Grinnell, and Dave Perkins. Our work consisted of determining concentrations of respiratory poison gases, mainly phosgene, under desert conditions. We had a very dusty lab, and it was extremely hot. We were very casual about gases. I remember going in a jeep to pick up a Dewar flask of phosgene, and riding bumping back holding the open flask at arm's length with the gas coming out.
>
> Some nitrogen dioxide was mixed with phosgene to make the clouds visible, so we wanted time lapse photos. I rigged up a windshield wiper with a variable voltage rheostat source from a battery. This was used to trip a movie camera at voltage dependent intervals. We got some good pictures that way.

AN IDYLLIC ISLAND IN CALIFORNIA

The field tests at Suffield in England were done on flat grass-covered ground. Yost particularly wanted NDRC to study flat and brush-covered desert, and also to study forested areas and beaches. Dickinson and others searched coastal southern California seeking lightly populated areas that might be cleared of people in order to carry tests with chemical weapons. They recommended Santa Cruz Island off the coast from Santa Barbara, California. They obtained permission from the owners for us to explore the island.

Professor Dickinson, I, and a car full of others, including members of Yost's group, drove along the Pacific ocean coast about a hundred miles to Santa Barbara, California, loaded our equipment on a commercial boat, and crossed the water for thirty miles to Santa Cruz Island. One of the boat crew told us that European wild boar had been introduced to the island to provide big-game sport hunting. With no natural enemies, the wild hogs had multiplied rapidly and become a major pest to the agricultural activities, and the vicious animals were a hazard to the few residents. The owner hired professional hunters and trappers to try to exterminate the hogs, but the wild hogs quickly learned how to stay just out of range of the hunters and to avoid the traps and poisons. We were further told that a disease, harmless to people, highly contagious and fatal to swine, was introduced, and it appeared that wild hogs were exterminated on the island, but recently survivors had been seen in the most westerly region. One former agricultural activity on the island was to raise Merino sheep for their high quality wool. This effort was dropped, but some of the sheep had escaped and roamed wild. The island is about twenty-five miles long (40 km) with an average width of about five miles. Four islands span a length of about eighty miles (129 km) along the Santa Barbara Channel: Anacapa (uninhabited), Santa Cruz, Santa Rosa, and San Miguel. The boat took us to Prisoners Harbor, which consisted of a long curved beach, one simple wooden pier, and an abandoned two-room house built of concrete blocks. We stored our provisions in the old house and pitched our tents on the beach. Professor Dickinson stayed only long enough for us to unload. He gave a date when he would return.

The island was privately owned, about ninety-five percent belonged to one owner and five percent belonged to the other. One dirt road led up to the substantial house of the owner of the larger portion, and another dirt road led up to the previously undeveloped smaller portion. The Navy had leased the smaller portion, which was said to have a commanding view of the Pacific Ocean in that region. They set up temporary buildings for the resident crew, for their telescopes, and a tall radio broadcasting tower. We were told by the operator of the boat that the Navy station there had spotted a Japanese submarine far off the coast during the previous week. I never heard any confirmation of that report.

We set up the micro-meteorological instruments just above the high-water line on the beach. Part of our job was to explore the area and provide a written report of its terrain and vegetation. It was only a two-man job to read the meters and write down the values so a couple of us would attend to the instruments and the others could hike, fish or swim (swimmers found the water to be cold). Around the harbor there was an unfamiliar and amazing forest of giant eucalyptus trees. A hike up the dirt road that went inland provided us with a view of the owner's house in the distance. We broke into groups to explore and write descriptions of the nearby land.

John Sullivan from Yost's group spent so much time at the beach and in the sun that he developed third degree sunburns, that is, blisters, over the tops of his feet. He fabricated sunshields out of cardboard, which resembled upside down swim fins.

Professor Dickinson did not return on the scheduled day. We had been profligate with the rations, and we found ourselves low in food. After a controlled dinner that night, someone caught a long large fish. We cleaned it and mounted it on a clean plank, which we suspended with wires in the concrete block house. We looked forward to a breakfast of this fine fish. Later that night, one member of our group called us in to see the fish, which glowed in the dark. Someone said this glow indicated bacterial activity, and we buried the fish. Later I read that this phosphorescence was natural for fish in this area because they fed on certain phosphorescent shrimp.

For some reason our gear included a .44 automatic revolver. John remembered seeing it in one of the drawers in our laboratory. Were we supposed to defend ourselves against Japanese invaders if their submarines landed off shore? Did we have this weapon to defend ourselves from wild hogs? During my free time that day, I took a long hike through the forest along animal trails. By some line of strange logic, I thought we needed extra food, and I took the pistol along with me. Eucalyptus trees grew well apart from each other, the ground was littered with long strips of bark that had peeled off the trees, there was scant underbrush, and I could see for long distances between these trees. There were other kinds of trees, singly and in groves, accompanied by heavy underbrush. Numerous wild flowers grew in one small green-grass meadow. I stopped to look closely at a spring of water that flowed out from under a large boulder into a clear pool, and, surrounded by green grass, the water trickled down the gully. Unfamiliar lilies blossomed around one side of the pool, and shrubs grew out of the gully bank. I sat on a small boulder to see if there were any fish or frogs in the pool. When I heard noises from sheep nearby, I assumed they were coming to get a drink from the pool. I climbed from the gully and sat still on a rock. Soon a flock of sheep appeared, obviously descendants of those that had gone wild. The handsome young lambs were uniformly covered with clean short white wool, including a wagging long white tail. Merino sheep had been bred to have long high-grade wool that adhered to the sheep until people sheared them. They were not adapted to run wild through underbrush and briars. Some wool of the adults hung down to the ground as dirty masses of hair, sticks, and filth, and many of the adults had large bare spots where the wool had been pulled out as the sheep pushed through low growing vegetation. The long tails of the sheep were especially matted and messy. A large dirty ram led the flock; he stopped between two trees fairly close to me, and cautiously looked in all directions. Perhaps I foolishly thought I could supply the camp with additional food, or perhaps I felt my sense of power. Here I had a loaded gun in my hand and there was a creature I could kill. I released the safety catch, aimed the gun at the ram's head, and pulled the trigger. The .44 pistol fired with a loud bang, and the recoil bent my arm and kicked the gun in my hand up over my head. The ram stamped his foot and rapidly looked around, but he did not leave his place. I took aim again, held the gun with both hands, and fired again. The recoil pushed both my arms back, and again forced the gun up over my head. The ram stamped his foot and snorted, and they all ran away. I immediately realized how foolish I had been. We had no knives and equipment to skin and clean an animal, and, anyhow, I had no way to get a carcass back to our camp. I sheepishly said nothing about shooting at sheep.

That afternoon, Dickinson returned in a speed boat. We loaded our gear and left the island.

* * *

Somewhat later on the mainland, Professor Dickinson and three army officers visited the owner of ninety-five percent of Santa Cruz Island. Being secretive about the details of their plan, the Army officers explored whether the owner would be willing to lease the island in support of national defense. Perhaps, he would have been willing for some defense purposes, but he recognized the insignia of the chemical warfare service on the visiting officers' buttons, and he firmly rejected the proposal, whatever it was. He did not want his island to be contaminated by large scale tests of chemical warfare agents.

RESEARCH IN CALIFORNIA DESERT

Professor Don Yost and Dr. Richard Dodson came back to Caltech in January 1943. According to the terms of their contract, they set up their own proving ground in the Mojave Desert on Rosamond Dry Lake, which Yost named El Rancho Grande. Yost's NDRC project had about eight people working on it; the four of us on Dickinson's project worked directly with them. To reach the desert research station from Pasadena, we drove up the dry west-facing slope of the San Gabriel Mountains, into and through a forest of tall pines and Douglas firs and low growing bushes, out of the mountains into the Mojave Desert, past the small villages of Palmdale and Lancaster, to the couple of buildings named Rosamond, where we turned right to the large flat Rosamond Dry Lake.

Yost's group released sulfur dioxide gas in the air and traced its path and dissipation by chemical analyses. Sulfur dioxide is highly irritating to breath and is mildly poisonous, compared to phosgene. We contributed to this work by making micro-meteorological measurements and observations. Our measurements included temperature and wind speed profiles, pictures of small smoke clouds, and use of the British gustiness meter. The floor of the dry lake bed consisted of tiles of salty cracked dried mud, almost devoid of plant life. There were low growing shrubs in some raised areas. We set up our main station with all of the instruments. To sample both the flat and partially vegetated areas, we had a satellite station where we made a limited number of measurements. We commuted between stations by bicycle. Eventually we added a sail to the bike for recreation. The desert, the instruments, and operation of a smoke pot are pictured in Figures 2.5.

Two members of Yost's group came to Caltech from the University of Arizona as graduate students. Yost wrote to the chairman of the chemistry department at Arizona in 1943 saying he had two excellent people from the department and wondered if there were any men in the class of '43 he could recommend. The chairman wrote back that there were only two of his graduates he would recommend, but one was already committed to attend UC graduate school and the other was a young woman. Rene Scott was hired and became an appreciated and valuable member of the group.

In the early stages of this work, we drove our Buick station wagon with all equipment inside or strapped on top from Pasadena to Rosamond Dry Lake, set up our apparatus, made measurements, packed up the equipment, and returned to Caltech. It was a long day's work. When Yost's group studied gas travel at night, Bob and I brought enough food for two

Figure 2.5.A. Rosamond Dry Lake, Meteorological Instruments, 1943.

Figure 2.5.B. Rosamond Dry Lake, Smoke Pot, 1943.
Photographs by Roscoe Dickinson, given to HSJ in 1943.

days, and camped with army cots and sleeping bags. In this way we got twenty-four hour samples of data at one hour intervals. During the night, one of us got up and read the meters every two hours, and as an act of fiendish glee slipped the alarm clock under the cot of the other, which was to ring an hour later to announce his turn to get up.

EGBERT

Professor Dickinson suggested we should reduce the size and complications of our gas analyzing system so it could be used as a portable device to measure war gases in the field. In the laboratory, our system had a twenty liter carboy of distilled water placed on a shelf one and a half meters high. This water flowed by gravity along the inner surface of a half-meter-high burette, where water scrubbed soluble gases out of the counter-flowing stream of air. The water then flowed through a cell that measured its electrical resistance and recorded it on an Esterline Angus meter. The flowing stream of air expanded from high pressure in a heavy steel tank. In the field, there would have to be a supply of distilled water, which was an essential ingredient in our system, and a pump would be needed to pull atmospheric air into our system. It appeared to be a difficult job to miniaturize this system.

John Otvos primarily carried out the miniaturization of our analytical system, but I made one contribution. Dickinson jokingly said I should use some feature of watering chickens to do our job. With that challenge, I played around with water and gas flows for a couple of days. After some trials, I took a long straight glass tube and attached an open T-tube at the upper end. I tilted the tube slightly and let water slowly flow into the upper end of the tube. By controlling the tilt and by applying suitable constrictions to the incoming and outgoing flow of water, I could get air to enter the open T-tube and be picked up by the down-flowing water. With more adjustments I could get the incoming air to break up into round bubbles, and the flow down the tube looked like a string of pearls. The flowing distilled water sucked in outside air, and the bubbles of air flowing down the tube would scrub out any soluble war gas in the air. This simple method eliminated the need for a pump to pull in outside air, and it replaced the heavy carboy of water with a compact two liter bulb of water. I made this minor discovery on October 2, 1942, according to my notebook.

Art Stosick developed the electronics. John replaced the long straight glass tube with a glass-tube spiral. He and Art designed and had the shop build a miniature cell in which to measure electrical resistance of water. The woodworking shop built a box that held the spiral tube in one compartment and held the electronics in a parallel compartment. A simple dry cell battery provide enough power to run the machine for several days. John placed a meter on the front panel, which read electrical resistance, and he provided two binding posts that could be used to hook the machine to an Esterline Angus recording meter. John named the new machine "Egbert." (Figure 2.6.).

SEARCH FOR FORESTED TEST SITE

Representatives from Dugway explored several parts of the country in an effort to find a remote forested site, which resembled the jungles of the south Pacific and where chemical warfare weapons could be tested. A heavy growth of pines in the Targhee National Forest of Idaho was tentatively selected, and during the summer of 1943 a Dugway Proving Ground expedition carried out tests there. Professor Malcolm Dole was in charge of the NDRC group, and a captain was in charge of the army group. Real chemical bombs were fired on the ground. (Noyes, page 321).

Arthur Pardee describes some of this work from the point of view of a recent graduate student:

> I think most of our group went to Targhee National forest, for a week or so, in the summer of 1943; the weather was good. We tested gas dispersal under forest conditions. It was technically pretty much like at Dugway. I did get a sense of the danger when a local forest ranger participated in one of our shoots. I was responsible for him, and he was quite worried about the gas which was old stuff for me by then. I was more concerned about getting behind a tree to avoid bomb fragments.

Pine forests were not much like the jungles of the southwest Pacific, and a team from Dugway Proving Ground searched for a more suitable area. They found a much better site in the Withlacoochee Soil Conservation Project near Bushnell, Florida. The Dugway Proving Ground Mobile Field Unit of the US Chemical Warfare Service was set up in the Fall of 1943. Malcolm Dole was to be in charge of the NDRC Division 10 unit, and all the NDRC group at Dugway moved to Florida. Captain Jake Nolen was the commanding officer. (Noyes, page 321).

A B C

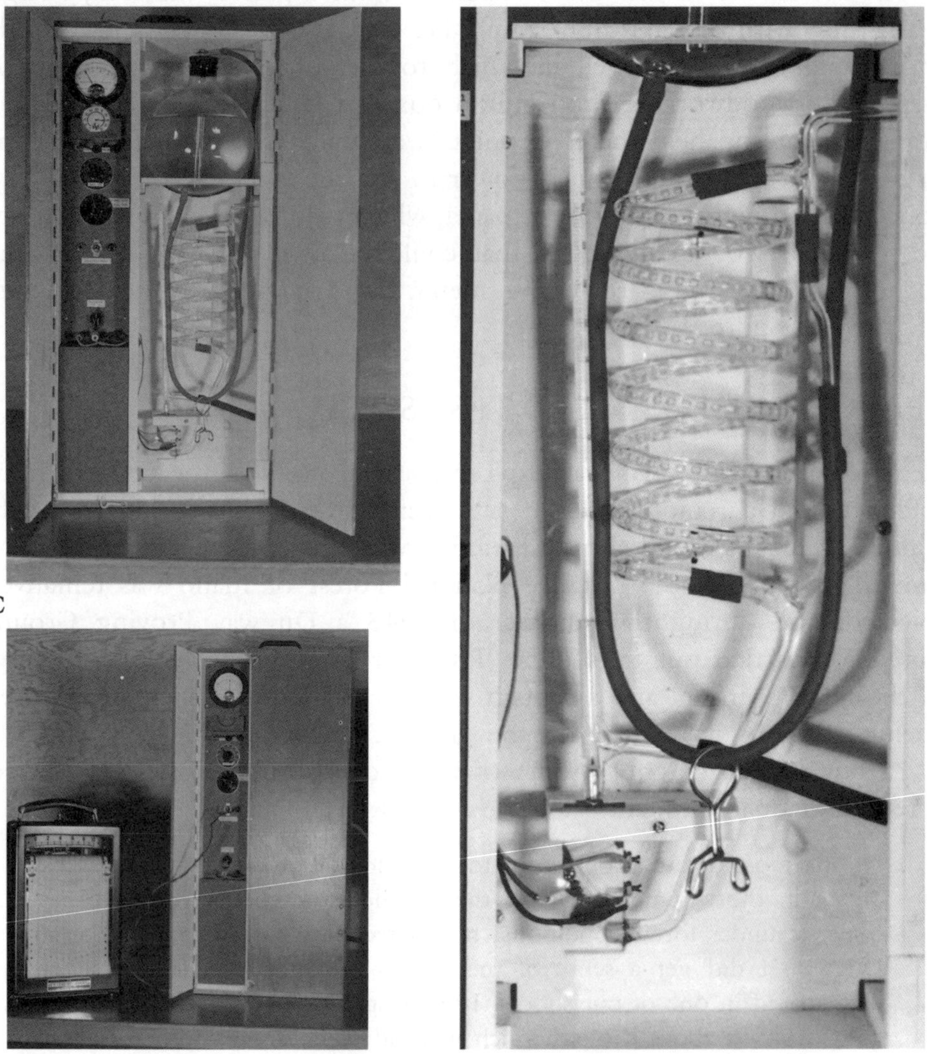

Egbert. Our portable instrument to measure gas concentrations in the field by the method of electrical resistance (conductance).
(A) The full instrument.
(B) The glass coil showing the bead of bubbles and the conductance cells at the bottom.
(C) The instrument hooked up to an Esterline Angus recording meter.

Figure 2.6. Photographs supplied by John Otvos.

* * *

Through August 1943 Professor Don Yost was in charge of all research of NDRC Division 10 as well as his group at Caltech, and he was highly important in planning the impending field tests. Effective September 1, 1943, he suddenly resigned from all NDRC work and donated all his NDRC equipment to Northwestern University. Some of his Caltech group transferred to other Division 10 groups, and some joined the Manhattan Project, which developed the nuclear bomb. W.A. Noyes, Jr. was administrative head of Division 10. He wondered if he had done something to offend Yost, and he wrote a long letter of analysis and apology. It appears that there was some dispute between Yost and others in NDRC, but I was unable to find the details. The cause may have been that Yost was desperately sick. He had a massive bacterial infection in his jaw bone, and doctors said such a bone infection was almost invariably fatal. He was in the hospital for months undergoing intravenous injection of penicillin. Yost was among the first whose life was saved by penicillin.

PANCAKING

Ken Pitzer and Sam Ruben, both instructors in chemistry at the University of California, were in charge of the NDRC chemical warfare group at Berkeley. After a short time, Pitzer went back East to work in a top secret war research project, but before he left, he made a major contribution to chemical warfare problems. During a trip to Dugway, Pitzer noticed that when a bomb on the ground was exploded, it quickly produced a hemisphere of gas, but then the upper part of the cloud collapsed toward the ground increasing the area of the cloud. This was later found to occur with all non-persistent gases, and it was termed "pancaking." After detonation, the liquid butane or liquid phosgene evaporated to form a gas, which cooled the air in the bomb cloud, making it heavier than the surrounding air, so it fell toward the ground. Pitzer worked out the mathematical explanation of the chemistry and physics of pancaking.

The purpose of the Berkeley project was to understand how gas clouds traveled and dissipated over various terrains and micro-meteorological conditions. They developed new instruments to make such studies. If an electrical current flows through a platinum wire, it may become red hot, and its temperature was easily measured. If air flowed over the hot wire, its temperature was lowered, and the faster the air flows the more the temperature is lowered. The Berkeley group measured wind speed by use of this "hot-wire anemometer." If air contains some butane, for example, the hot platinum wire burns the butane, and the heat of combustion causes the wire to get hotter. Ruben and his associates placed a hot-wire anemometer inside a tube and aspirated air over it at a constant rate, and this system gave a good measure of trace amounts of butane in air. At night in the California central valley they found exceedingly high temperature inversions and their butane source built up surprisingly high concentrations of butane. They especially studied katabatic winds down the lower slopes of Mount Shasta.

At an earlier time, Don Yost of Caltech had suggested adding a small amount of nitrogen dioxide (a red gas) to a colorless gas cloud to make it visible. Although nitrogen dioxide is poisonous, a few percent of it dissolved in butane would give a visible cloud much less toxic than phosgene, and butane concentrations could be easily measured by simple hot-wire anemometers. The NDRC group at Berkeley exploded a small number of bombs filled with butane containing some nitrogen dioxide, and they obtained important information about gas-cloud dissipation. As a result of this work, the Army often added five percent nitrogen dioxide to phosgene to make its clouds visible.

A TRIP TO MOUNT SHASTA

Since Malcolm Dole was to be the Technical Director in charge of all the NDRC groups in Bushnell, Florida, he visited each laboratory that was going to send personnel to Bushnell. In September 1943, he took an airplane to the Los Angeles airport, rented a car, and drove to Pasadena. He seemed to be substantially younger than Dickinson, perhaps about thirty-five. We showed him our meteorological instruments in the cow pasture to the east of Caltech, including the stack of windvanes. He examined the photographs we had taken of smoke-pot clouds during the windy turbulent midday and during strong temperature inversions at sunset. He admired our new portable field analytical instrument, Egbert, and recommended that we have our shop make as many as we could carry with us to Florida next month.

Professor Dole wanted next to visit the NDRC project at the University of California at Berkeley, which was directed by Sam Ruben and was then operating at Mount Shasta. He and I drove from Pasadena to Mount Shasta in late September. We left before sunup, drove a few miles down Highway 66, turned onto Highway 99, and stayed on Highway 99 for the thousand kilometer [600 miles] trip to Mount Shasta. First we climbed up the local mountain range and down the grapevine grade to Bakersfield, where, as in Los Angeles, there were many active oil wells. We drove up the center of the great San Joaquin/Sacramento Valley. The Sierra Nevada range was visible to the east and coast range mountains could be seen to the west. Almost level fertile soil and thriving crops stretched from one mountain range to the other, for a width of about fifty miles. In the southern part of the central valley, Highway 99 had four lanes, two going north and two going south. The roadways were separated by a strip of land with large oleander bushes, which were loaded with blossoms, some white and some brightly colored. During the war, the national speed limit was forty-five miles per hour, but we drove much faster than that. The trip took all day, with occasional stops for gasoline or food. We plodded through Sacramento, through Williams and Willows, and through Red Bluff and Redding. We could see the snow covered mountain long before we arrived there. Late in the afternoon we reached Dunsmuir and turned up a narrow mountain road to Mount Shasta.

There was still some daylight when we reached the group from the University of California. There were about a half dozen people there, including the wives of two of the team. However, I remember the names of only two. Bill Gwinn was the senior member of the group, and Dole talked primarily with him. John Thomas was the junior member of the group, and I talked primarily with him. We set up our cots and sleeping bags, and cooked our supper on the still-burning campfire of the Berkeley group. John showed me the men's toilet, which was a slit trench with dirt piled up behind the trench, with a shovel stuck in the loose dirt and a roll of toilet paper stuck on the handle of the shovel.

John Thomas told me a great deal about the discoveries they had made and the adventures they had (see Chapter 3). He said I was unlucky not to have come the day before. Their leader was Sam Ruben, but he had left camp early that morning. John praised Sam highly for his intellectual brilliance, his ebullient personality, and his sense of humor. He was returning to Berkeley to calibrate their instrument to measure phosgene for use in an upcoming experiment on Stinson Beach near San Francisco and for use in the expedition to Bushnell, Florida. John repeated that I would have greatly enjoyed talking with Sam.

I was especially impressed by one of John's stories:

At Berkeley, Kenneth Pitzer and Samuel Ruben recommended use of non-toxic butane instead of highly toxic phosgene for the field studies of gas cloud dissipation. Phosgene is poisonous, dangerous to handle, and difficult to analyze in the field. Butane has similar thermodynamic properties and would be transported by air in the same way as phosgene. Sam Ruben and Bill Gwinn had developed a simple, ingenious, practical way to measure and record trace amounts of butane in air. In the late summer and early fall of 1943 Ruben and his crew found interesting micro-meteorological features at the Mount Shasta site and, using butane, carried out extensive studies of air motions in the forest. They felt that they had demonstrated the feasibility of using butane to simulate phosgene, but army representatives would have nothing to do with substituting the "real thing" by another chemical. The Berkeley scientists lost this argument with the army, and field tests with phosgene were planned for Mount Shasta.

There was a small town four miles (6.4 km) away at the foot of the sloping terrain that stretched out below the test site. Sam Ruben made some calculations based on their recent results with butane and warned that the town might be in danger of being poisoned by phosgene.

The army representative said that the town was four miles away, and war gases don't travel that far. The Berkeley group mentioned the slope, the terrain, and katabatic winds, but the army was unconvinced. Ruben and Gwinn came up with another idea and went ahead with their proposed novel experiment.

On a cold evening with clear sky conditions, Ruben picked Tom Norris to take a five gallon (nineteen liter) can of mixed mercaptans, that is, pure essence of skunk stink, and pour it out on the ground at the area where the army wanted to test multiple large phosgene bombs. On foot and by car, all members of the project went over roads and trails between the test site and the small town. They recorded the time, location, and degree to which they smelled skunk odor. Katabatic wind picked up the mercaptans, and the river of cold air slowly flowing down the slope carried the odor with it. Sniffing graduate students mapped the flow of the gas from the proposed test site. The residents of the town below probably thought the skunk situation was the worst it had ever been. They probably never realized that the terrible odor may have saved their lives, because this demonstration caused the phosgene tests at Mount Shasta to be canceled.

As we were ready to leave in mid morning of the next day, John Thomas strongly recommended that we look up Sam Ruben during our visit at Berkeley and hear about his plans for Florida, but Professor Dole did not choose to visit Berkeley. We drove straight to San Francisco. As we approached San Francisco, Dole spoke excitedly about the Golden Gate, the Barbary Coast, and Columbus Avenue. We stopped briefly on our way across the Bay Bridge and admired the view of the city. We checked into a hotel. Dole asked me to read a long report and left me there, and he went out on the town.

* * *

In October and early November, 1943, we spent full time in preparing our instruments to join the Dugway Proving Ground Mobile Field Unit of the U.S. Chemical Warfare Service in Bushnell, Florida.

* * *

In his 1948 book, W.A. Noyes said about Egbert on page 325:

> One of the most useful recorders was developed by Dr. Dickinson's group at the California Institute of Technology. The gas is entrained in bubbles in a downward-flowing absorbent liquid and the concentration of the resulting conducting solution is measured by a suitable conducting cell and circuit on a milliammeter or on a continuous recording milliammeter. More than a hundred of these instruments were manufactured and they received widespread use at Dugway Proving Ground, at Bushnell, Florida, at San Jose Island, and in various miscellaneous experiments.

CHAPTER 3

SAM

This chapter is a brief biography of a remarkable scientist, Samuel Ruben. In 1943, Ruben was in charge of a National Defense Research Committee (NDRC) research project at Berkeley, which concerned chemical warfare work. The research done by this project paralleled in many ways the work done at Caltech as described in Chapter 2. John Thomas told me that I was unfortunate for not meeting Sam when I visited their research site on the side of Mount Shasta in September 1943. Sam had returned to Berkeley twelve hours before I arrived. Prior to his chemical warfare work, he was the co-discoverer of carbon fourteen, and he made outstanding discoveries in the field of photosynthesis with artificial radioactivity produced by the new Berkeley cyclotrons. By age 29 he was world famous for his research in both physics and biology. With a slight stretching of the term, he was among the first molecular biologists. As background, I present here some of Sam Ruben's ancestry and family, and of science at Berkeley in the 1930s and early 1940s. Since I was not present during any part of the events of this chapter, I make heavy use of direct quotations from those who were there. Most of the material in Chapter 3 was given to me by Sam's son George Ruben.

A comparison of Chapter 1 and Chapter 3 shows that Sam Ruben and I were similar in a small number of ways, but we were opposite with respect to strength, health, scientific knowledge, and, especially, luck.

Background

Sam Ruben's grandfather, Jacob Rubinstein, was born about 1864 in Ostralenka, Poland, near the Russia-Poland border. In his spare time he was a Talmudic Scholar. He worked at various occupations, as a clerk, a buyer and seller of goods, and as a smuggler. For example, he purchased silks and tobacco in Danzig and smuggled them into Ostralenka. For forty-four years of his life he lived in Poland, and in 1908 he immigrated to New York. In 1910 he became a naturalized American citizen. He died on April 1938 in Bronx, NY, at seventy-three years of age.

Sam's grandmother Shaine Yitke Shapiro was born about 1864 in Ostralenka, Shaine and Jacob married in 1881, and they immigrated to New York in 1908, where they lived in a tenement at 1659 Washington Avenue, Bronx, New York. In the 1920 census, Shaine was listed as a housewife with "no occupation," and Jacob's occupation was described as "peddler, rags, wage worker." Both Jacob and Shaine spoke Yiddish as their native language, and they could not read, write, or speak English. Shaine died January 1932 in Bronx, New York, at sixty-seven years of age.

Sam's father, Herschel Rubinstein, was born in Ostralenka, Poland on May 5, 1882. In Poland, he studied to be a rabbi but did not complete this program. Herschel became an expert capmaker. At this time many men in central Europe wore caps, most immigrant workers in New York wore caps, and Herschel provided a comfortable living for himself and his family by making and selling caps. Sam's mother, Freida Penn, was born on April 6, 1883 in Lithuania, Poland. The Census of 1920 states that Freda Rubinstein was age 37, immigrated to USA in 1910, was able to read, write, and speak English, and that she had no occupation.

Herschel Rubinstein and Freida Penn were married at about 1905. Their wedding portrait in Poland is shown on Figure 3.1.A. Herschel and his family immigrated to the United States in 1910, they spent some time in New York, and then they moved to San Francisco and resided at 1686 Grove Street. Herschel became a naturalized citizen in 1926. The Certificate of Naturalization stated that he was forty-four years of age, five feet and three inches [66cm] tall, white complexion, gray eyes, and dark brown hair. By court order dated March 13, 1930, his name was officially changed to Harry Ruben.

YOUNG SAMUEL RUBEN

Samuel Ruben was born in San Francisco on November 5, 1913. Figure 3.1.B is a picture of Sam at age four, his older sister Ida at age nine, and his two year old sister Mae.

The warmth of a fine wool cap was rarely needed in the bland San Francisco climate, and wearing a cap there became rare, even among central European immigrants. Harry Ruben apprenticed himself as a carpenter, and his family survived on the wages he made as a carpenter and from the occasional sale of a cap. Even before the depression, Harry Ruben was a poor man. With the onset of the depression, construction of houses almost completely ceased, and jobs for carpenters vanished.

During the period of poverty in San Francisco, young Sam had one golden adventure. The famous boxer, Jack Dempsey (1895–1983), lived less than a block away from the Ruben family. Dempsey fought sixty-nine times, knocking out his opponent forty-seven times. As a child, Sam had a strong case of hero worship for his distinguished neighbor. In 1919, Dempsey won the world heavyweight championship from American Jess Willard, and in 1921, when Sam was eight years old, Dempsey successfully defended his title against Georges Charpentier from France. Sam followed the boxer's triumphs with enthusiasm, as he retained his championship in a 1923 bout with Luis Angel Firpo from Argentina. In 1925, the twelve year old Sam must have felt for the great fighter when he lost the world heavyweight championship to Gene Tunney, and again two years later when Dempsey lost a second time to Tunney in the famous "Battle of the Long Count."

We do not know the details of Jack's relation to neighborhood boys in San Francisco at that time, but what we do know is consistent with the following: The boys in that part of town had a boxing club, which Jack Dempsey occasionally visited, giving demonstrations with the slugging bag and lessons to some of the boys. Sam Ruben stood out as a boxer and as a person, and Jack Dempsey gave Sam a full page photograph of himself inscribed, "To a boy frend Sam Rubin," and signed "Jack Dempsey." (Figure 3.1.C).

IDA RUBEN

Harry Ruben returned to New York because he could not earn a living in San Francisco making caps and doing small jobs. Meanwhile, Ida had a job in Berkeley. Freida, her son Sam, and her daughter Mae moved to Berkeley, where Ida supported them. Sam worked part time as a salesman at Call Me Joe's clothing store in Berkeley.

Ida Rubinstein was born in Poland April 17, 1908. Ida graduated, January 1926, from the High School of Commerce in San Francisco, and she graduated from the University of California, December 1932. Her Major Field of study was German. Her Minor Fields of studies were English, History, Spanish, and Education. In a general sense of the word, she was the strongest person in the Ruben family. Figure 3.1.D shows Ida as a young woman. Ida married Russell Millard Slee. Russell and Ida lived at 1414 Hawthorn Terrace, Berkeley, California for forty five years.

Ida helped many members of the family. After the death of Ida's mother, Ida's father lived with her. She gave Mae and Sam financial help, so that they both could attend the University of California. Ida's twenty year old brother Sam appraised her in a letter he wrote on July 18, 1934:

(a)

Mr. and Mrs. Harry Ruben or
Herschel Rubinstein and Freida Penn.,
Sam's parents

(b)

Sam, Mae, and Ida Ruben

(c)

Jack Dempsey

(d)

Ida Ruben

Photographs 3.1. Supplied by Dr. George Ruben.

Sam, High School Senior

Basketball Star, Right Rear.

College Senior. Yosemite: Hikers Lodge

'Nevada Falls, Yosemite

Photographs 3.2. Supplied by Dr. George Ruben.

Mother has been confined to bed for 6 days—Ida and Mae do all the cooking etc. Gosh can Ida cook! She can make the best meals I've ever tasted—I'm not hard to please but nevertheless I've eaten a lot of meals since was born. (Damme if that isn't so!) The fellow that marries her will sure be lucky. But she says she is going to be an old maid and if I will remain a bachelor she will take care of me and cook for me—and its almost worth considering. I just wrote a letter to Ida and I spelled all the words wrong on purpose. Will she be mad—and will I be glad. Goody goody—sweet cakes.

Mae Ruben was born in San Francisco, July 1915. After she graduated from the University of California at Berkeley, Mae was a bookkeeper and writer. She worked for Standard Oil in Richmond, California, and the Albany Times. Mae lived with her sister-in-law, Helena, from 1943 to 1948. Later, she had her own apartment. When she became ill from emphysema, she lived for a time with Ida, and as her illness progressed moved into the Jewish Home for Parents in Oakland, California, about a year prior to her death in 1982.

ATHLETIC SAM IN THE SIERRA NEVADA MOUNTAINS

Figure 3.2 shows Sam as a high school senior, as a basketball player, and as a college senior in the mountains of Yosemite. Throughout high school Sam was a star basketball player. During his college years, he camped and hiked for long distances in the Sierra Nevada mountains, carrying a sixty pound backpack. He describes himself at age twenty and his mountain hiking experiences in letters to Helena West:

Camp Curry, 6/15/34
Yosemite National Park,
California

Dearest Helena:

Just got in Camp Curry from Merced Lake. Was I glad to get your 2 letters! I ran the last 2 miles down the valley just in anticipation—I almost killed R, J and B with astonishment and the other people on the path by bumping into them.

I got into Camp first again today. Am I tired! But I feel very rough and tough—not having seen but about 8 people all week (and only one woman who was the wife of the caretaker of the Hikers Camp—she took a shine to me—actually gave me a loaf of bread, and wanted to give candy as well but I would rather buy the candy—the bread was great).

Roy and Johnnie left for Berk immediately while Beardsley is fixing the car. You ought to see us now! I don't think you'd speak to me.

5 days growth—haven't combed my hair except with my fingers, keds are patched with adhesive, etc.

It rained, hailed and (even snowed a little just above us) but I didn't give a darn. Nights were freezing cold but I'd always be the 1st out to make the fire—I'd go around in my pajamas and when the sun would come out I'd even take them off and put on my shorts.

I can hardly wait until I get to home and you.

Love, Sammy

HELENA WEST AND FAMILY

Sam's letters are addressed to Helena. George Hazzard West was born 1882 at Philadelphia, Pennsylvania. He attended the University of Pennsylvania (1900–1904), obtaining the Bachelor of Science degree in Chemical Engineering with Senior Honors. He was elected member of the Society of Sigma Xi, 1904, as was his son-in-law, Sam Ruben, and his grandson, George C. Ruben. (both Sam and George were elected to Phi Beta Kappa). His profession involved the chemical aspects of mining, especially of copper. George West and Rosina Collins were married, and they had two daughters, Helena and Maida. The family moved to Berkeley, California in 1919 when Maida West was 3 and Helena West was 6 years old. George West and Rosina West bought the house at 651 Vincente Avenue, Berkeley, in 1927, where Helena lived until 1999.

University of California Cyclotrons[#]

The University of California cyclotrons were major features in Sam Ruben's education and in his research at Berkeley. In 1928 at age twenty-seven, Ernest Orlando Lawrence left Yale University for the University of California at Berkeley to become an Associate Professor of Physics, and at age twenty-nine he was promoted to Full Professor. Also in 1928, Robert Oppenheimer joined the physics department at Berkeley as an assistant professor. By that time, a great deal was known about the structure of atoms: there is an extremely small dense central nucleus and a diffuse cloud of electrons distributed about the nucleus. Physicists, then, were strongly interested in developing methods to explore the nature of the nucleus, especially at Cambridge, England, but also at top physics departments in other European countries and

---

[#]*An American Genius: The Life of Ernest Orlando Lawrence*, by Herbert Childs, E. P. Dutton & Co., Inc. New York, 1968.

in the United States. Such exploration needed machines that could accelerate charged particles with a million volts or more. In terms of strength of materials, such high voltages are extremely difficult to handle and use. In his second year at Berkeley (1929), Lawrence developed the concept of the cyclotron, which could produce the needed high-energy particles by multiple successive pulses of much lower voltages. The machine required a large powerful magnet, a half-circular disk made in the shape of a D and its mirror image, a high vacuum system, a radio-frequency oscillator, and other electrical supplies. Many top ranking physicists regarded it as a clever idea but doubted that it could be put into actual operation. Using theory and trial-and-error, Lawrence and his associates developed first a small working cyclotron and then larger ones. In January 1931, Lawrence and associates made the first demonstration of a working cyclotron, which was four inches in diameter (10 cm) and produced effects equivalent to 80,000 volts. This machine did not give the needed million volts, but the eleven inch (twenty-eight cm) cyclotron finished in August 1931 did provide information of interest to physicists.

Over the years, Lawrence went repeatedly to President Robert Sproul with a lucrative outside offer in his hand, and said that if the University of California did not match the offer by raising funds to build the next large cyclotron, Lawrence would be forced to leave. Sproul raised large sums of money for this purpose. The succession of cyclotrons at Berkeley is given by this chart:

Diameter		**Magnet**	**Electron**	**Date**	**Location**
Inch	**cm**	**weight/ton**	**volts**		
4	10		80,000	Jan.1931	Various
11	28		1,100,000	Aug. 1931	LeConte Hall
27.5	70	80	5,000,000	1933	Rad Lab-Campus
37.5	95		8,000,000	Aug. 1937	Rad Lab-Campus
60	152	200	32,000,000	June. 1939	Crocker Lab
184	467	4000	200,000,000	Nov. 1946	Rad Lab “Hill”
184	467		1,150,000,000	1956	Rad Lab “Hill”

The 27.5 inch (70 cm) cyclotron of 1933 and the 37.5 (95 cm) cyclotron of 1937 opened new territory for physicists, especially the production of many new artificial radioactive species. The cyclotron produced intense beams of high-energy charged particles, and when a beam hit a solid target, there was in some cases “transmutation of the elements,” to use an old phrase from alchemy. A more modern, yet incomplete statement is: When a high-energy beam particle hits a nucleus, it might knock a particle out of the nucleus, or it might donate a part of itself to the nucleus; the ejected or

added particle might be charged or not charged. If the mass of the nucleus is changed, it is a different isotope; if the charge of the nucleus is changed, one element has been turned into another. An early experiment showed that gold could be converted to lead, slowly and expensively. In a large number of cases the changed nucleus was radioactive, that is, it spontaneously ejected a particle or high energy radiation or both. Some radioactive nuclei decay within a small fraction of a second, and some decay over a period of billions of years.

The cyclotrons at Berkeley produced a large number of new radioactive species. Physicists were especially interested in the nature of the new nuclei and developed theories of how they worked. Chemists were useful in separating a desired radioactive species from a complex mixture of many species, and chemists used radioactive tracers to follow complex chemical reactions. Biologists were especially interested in using radioactive substances as tracers for studying transport and reactions within living animals and plants. Berkeley was the only place in the world with cyclotrons, many scientists from many places visited Berkeley. Lawrence sent samples of special radioactive species to other laboratories. The elements carbon (C), oxygen (O), and hydrogen (H) are essential to all living species, and radioactive tracers for these species were especially needed, but early attempts to make any of these in a form useful for biologists were not successful.

On November 9, 1939, Ernest O. Lawrence received word that he had been awarded the Nobel Prize in Physics, but because of the war, it would not be possible for Lawrence to go to Sweden to accept the award.

Sam Ruben's long-term partner, Martin Kamen[§] wrote the following:

> On February 29, 1940, there was a momentous ceremony in Wheeler Hall, where the Swedish Consul presented the Nobel Prize to E. O. L ... The chairman of the Physics Department, R.T. Birge, gave the presentation address ... He spoke of the great importance of radioactive isotopes as tracers in biology and possibly as therapeutic agents. Then in a dramatic gesture wholly atypical of him, he stepped back, raised his arm, and portentiously announced, 'I now ... have the privilege of making a first announcement of very great importance. This news is less than twenty-four hours old and hence is real news. Now, Dr. S. Ruben, instructor in chemistry, and Dr. M. D. Kamen, research associate in the Radiation Laboratory, have found by means of the cyclotron, a new radioactive form of carbon, probably of mass fourteen and average life of the order

---

[§]Martin D. Kamen, *Radiant Science, Dark Politics, A Memoir of the Nuclear Age*, University of California Press, Berkeley and Los Angeles, California, 1985.

> of magnitude of several years. On the basis of its potential usefulness, this is certainly much the most important radioactive substance that has yet been created.'

The author of E.O. Lawrence's biography† made this assessment:

> Carbon 14, most useful of all artificial isotopes, had been discovered. It is this carbon isotope which dates antiquities, and has become of major importance in biological studies.

The occasion and timing of Professor Birge's announcement implicitly prophesied that Samuel Ruben and Martin Kamen would, in the future, be strong candidates for the Nobel Prize for this discovery. At age twenty-six, Sam Ruben and Martin Kamen were world-famous in physics.

"Dearest Helena"

During 1934 and 1935, Sam Ruben sent a series of handwritten letters to Helena West. Helena replied, but we do not have any of her letters. Sam's letters given in previous sections express his youthful exuberance and his great physical strength. I include excerpts from some other letters here with no change to spelling and punctuation. Copies of all of the letters are in the Sam Ruben Archives, Bancroft Library, University of California, Berkeley, CA, 94720.

In innocent arrogance, twenty year old Sam wrote that Helena, who was "just a girl," should formulate a philosophy of life. The philosophy Sam advocates was Sam's philosophy for himself. From Sam's subsequent letters, it appears that Helena stood up for herself, to Sam's dismay, and over a period of time they squabbled furiously. However, Sam began to respect Helena's independence, and eventually Helena gently forced him to modify his stand and to say that he accepted her viewpoint, or at least part of it. In the end, love conquered (almost) all of Sam's arrogance toward Helena.

In the middle 1930's some California people were anti-Semitic in various degrees. From Sam's letters, it appears that Helena made statements that reflected the then current attitude of many people. Perhaps someone had made these statements to Helena in an effort to prevent her from marrying Sam. In some places Sam affirmed his pride in being Jewish, but in other places he was defensive and almost apologetic. In other letters Sam himself showed lack of sensitivity toward African Americans.

---

†Herbert Childs, *An American Genius: The Life of Ernest Orlando Lawrence*, E. P. Dutton & Co., Inc. New York 1968.

As an undergraduate student, Sam achieved strong recognition from his professors. Helena was Sam's one love, but the letters also express his affection for his many friends. Sam was extremely devoted to his work, chemistry and physics. The letters have extensive samples of Sam's whimsy. I have omitted strongly personal statements by Sam.

SAM'S PHILOSOPHY AT AGE 20
[No envelope, pencil on yellow paper, hard to read]
Sunday morning, July 8, 1934.

Dear Helena:

There are some things that have been burning my mind ever since I got up this morning. In regards to last night—let's please forget it as to actually what happened—but not it's significance. Now I realize that one of the things I want to do is to make you happy. The best way I know is to try and suggest a philosophy for you to follow which will make your life as I think you would want it to be.

You think a lot for a girl—but you only do it half way. Either do it 100% or leave your future in the hands of events. But you can shape <u>everything</u> to allow you your ambitions. Honestly—Helena—it is my belief that there is nothing in the world you <u>can't</u> do if you want to, and try hard and long enough. (of course being reasonable as well as logical in your demands). I feel that way—I have felt that way since I've entered college. But—believe me—there is nothing like the feeling that possesses you when you've done it. <u>Achievement</u>—that word means a lot to me. The feeling of having accomplished a worthwhile thing is unequaled. It is the biggest thing in my life—I think it always will be. When you're all through you've got something to show for your time. <u>Something</u>, not dollars and cents—not physical happiness—it is the supreme happiness—better than anything else because you've done it—sweated for it—sacrificed—slaved. And it is worth it.

It justifys my existence. What else is there to live for? If it wasn't for Chemistry I often feel my life was futile. When I die I want to have something to show for all the years.

What I want you to do is to adopt a philosophical view on your life. Not all at once, but it will grow on you if you think about it long enough. It doesn't come in a day, a week or a year but gradually it will possess you—and make you happy. The <u>mind</u> is the <u>basis</u> of <u>everything</u>—your <u>happiness</u>, your <u>sorrow</u>. Mental processes control your life. Utilize them—Make them do the things they should.

SAM DISLIKED WORKING AS A CLERK IN A CLOTHING STORE.

July 8, 1934.

Hell in 1 1/2 yrs I'll know whether I'll be something or just a dispenser of sox to broken down ladies who insist on matching the sox to the man's eyes. I hope my patience will last that long. What an occupation for a man. I'd rather be a Communist or a Red than a clerk. Well it will all come out in the wash! 1 1/2 yrs. isn't a long time.

THEY SQUABBLED FURIOUSLY

Envelope: Aug. 10, 1934
S. Ruben,
2415 Dwight,
Berkeley, Cal.,

To:
Miss Helena West,
651 Vincente Ave.,
Berkeley, Cal.

I just finished talking to you 5 min. ago so I'm still in the same mood. You're just a girl that thinks she is a woman—knows human nature 100%. You know a certain type of boy—and you think you know everything. I've let you tell me things that I would not take from anybody else...

The whole reason why I've blown so completely up is because to go thru the mental anguish—and I mean it—that I did—for <u>you</u> darn it—for <u>you</u>—not for myself—and to have the whole thing scorned and scuffed so completely without the effort being made to understand.

Well the hell with it all—you can forget you ever knew me.

BUT SOON THEY MADE UP

Envelope: Nov. 5, 1934.
[Sam's birthday]

From:
2415 Dwight,
Berkeley,
Calif.,

Darling Helena:

This is not going to be a crazy epistle. The fact is that here it is only Monday afternoon.—and you are noticeably absent.

But listen—if you tell anyone about this letter or read them its contents—or the idea behind it—I would not appreciate it. Because only love-sick sops do this sort of thing—or poets or writers. And I'm not a love sick sop! In fact I'm just an ordinary common everyday specimen of a dumb college student who likes a pretty girl, and don't you ever tell anybody that much either!

Thanks for your exceptionally good candied walnuts. They were very good. I liked them. So did my family whom I let taste them. Mother made a birthday cake—but it didn't taste anywhere as good as your walnuts. So there—you have a brilliant reputation with my stomack.

I'm pretty happy because I got my 102A mid-term [examination] back—but my 100 is beckoning me with a "come here you rascal" look in its eye.

Love By "Drunk, happy, and healthy and 21."

SAM'S RESPONSE TO ANTI-SEMITIC STATEMENTS TRANSMITTED BY HELENA

Envelope: April 8, 1935

From: 2415 Dwight,
Berkeley,
Calif.,

Dear Helena:

Fight—I shall always—for what I believe is right. I have no false pride for the Jewish race merely because I am a member. I shall not fight other's battles—but my own. I do carry my head high—but not so high as I can't see the faults, etc. of our race. Greed, money, wealth, etc.—they believe is the goal. For that I am ashamed. The fact that they contain men such as Einstein, Brandeis, Cardozo, etc. is nothing to crow about. Every race has its great men—just a question of probability. The point you make about the Jew invariably walking off with the spoils—I maintain is untrue. You cannot make that generalization—it may be true in certain cases—but not by a long shot—the majority. That is creed of all anti-Semites. True—I fought much as a small boy—but only when an unjust statement was made. But pig-headed I may be—but never to the fact that the Jews are supposed to be superior in that way—there are high-minded members—who possess all the ideals that high-minded Christians do.

I firmly believe that for our race to want to be a separate and distinct entity is wrong. I believe in tolerance, a broad-minded gradual diffusion—which is inevitable. I believe in inter-marriage.

I intend to live my own life, seek and find happiness that I'm a Jew. I've no regrets. I consider it neither an asset nor a liability. I don't consider religion essential—only for the morals it stands for. For that reason I minimize the whole thing. I may change my mind—I may have to—however that seems to be far in the distant future.

You have judged the Jewish race by the actions of the minority—at least that is my opinion.

All this as you have realized is an answer—point for point—to your letter.

I think I shall marry—but not to a member of the Jewish race. Irregardless of how stubborn I may be—I shall try, with all my heart and soul, to make her happy. Because in making her happy—I shall derive happiness.

To say it was the cessation of a mere infatuation is not true—it was more than that—it was love.

Please don't trouble yourself to answer this letter. I would rather this would be the finis. You were right—I blundered. What more need be said?

Love, Sam

INSENSITIVITY TO AFRICAN AMERICANS

Envelope: June 7, 1935,
Sund. eve 9:45

From:
2415 Dwight,
Berkeley,
Calif.

To whom it may concern:
(The sweetest girl in he world—I hope)

I just got back from the U.A.—saw a good picture—Oil for the Lamps of China—and a very screwey one—College Scandals. C.S. had lots of Chemistry profs—etc—and sweet girl chem students. I saw a girl in one of the scenes that reminded me of you—she was black—tall and scraggly—chewed terbakko—and sed "Holy smoke!"

Deane and I played ball—went down to the Y—had a sunbath, and a swim. Settled the affairs of the nation, kissed the Negro poster goodbye and came home. Created a famine at Dwight Way by eating 82 courses (not corsets)—went to the campus and read and slept. went to the show—came out and took a long walk.

Gee darling, I kept thinking of you all day. In the show—I closed my eyes and pictured you—talking to me. The plot of O.f.t.L. of C. was good—but somehow it increased my yearning for you. You see the whole theme and thesis was a man's life and happiness revolves about just two things

1. The woman he loves and who loves him.
2. His work

And darling the night is wonderful. It is cool and crisp. It is very clear and the sky is a deep deep blue—the stars are few—but bright. The moon is small but clear—it seems to repeat over and over again to me—"where is your Helena?" Oh—dear, everthing is so grand—everything smells so nice—it just makes your blood tingle—one deep breath after another.

Gee you wouldn't recognize me now—I'm as brown as 8 berrys. With my kinky hair—do I make a swell nigger! I changed my name to black Sambo. I bought myself a shoe shine pahla in Oakland—Yah suh! I'll suah can shine youah shoes—ma'm.

Please write soon!

With all my love
Black Sambo

MR. AND MRS. SAM RUBEN AND FAMILY

In the middle of Sam's last semester as an undergraduate student, on September 28, 1935, Helena West and Samuel Ruben were married. The newly wedded pair is shown in Figure 3.3.A.

In May 1935 Helena Gertrude West graduated from the University of California with the Bachelor of Arts degree in Chemistry. On December 20, 1935 Sam graduated from the University of California with the Bachelor of Science degree in Chemistry with Highest Honors. In January 1937 Helena Ruben received the Certificate for Secretarial Training. In May 1938 Sam received the degree Doctor of Philosophy in Chemistry. Sam joined the faculty of the University of California as Instructor of Chemistry in August 1938, and he was promoted to be Assistant Professor of Chemistry in August 1943.

After Sam and Helena were married, they moved in with Helena's parents at their home on 651 Vincente Avenue in Berkeley and lived with her parents for one year. They moved to a cottage behind a big house before Dana was born (November 11, 1938) and then moved to Colusa Ave. for about one

Helena and Sam, September 27, 1935

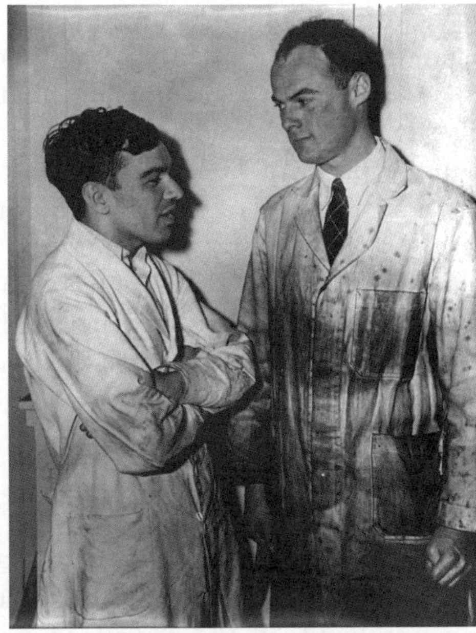

Martin Kamen and Sam Ruben, in the laboratory

Martin Kamen **and 37.5 inch cyclotron**

Sam Ruben **Research on Photosynthesis**

Photographs 3.3. Supplied by Dr. George Ruben.

year, where George Collins Ruben was born (April 29, 1941). They bought a house where Constance Mae Ruben was born (June 18, 1943). Figures 3.4.A & B show Sam holding his two and a half year old son Dana on his shoulder and holding his two month son George in his arms. The last photograph is of Helena with the three children, Figure 3.4.C.

In Sam's letter of June 7, 1935, he wrote "... a man's life and happiness revolves about just two things: 1. The woman he loves and who loves him. 2. His work." Undoubtedly, in intensity and depth of affection, Sam loved his family more than he did his work, but measured by time spent on evenings and weekends, his work came first, according to Helena (1997).

Sam's Brilliant Research Career in Science

Glenn Seaborg kept a daily diary since age fifteen. Seaborg entered Berkeley as a graduate student in the fall of 1934 and obtained his Ph.D. in 1937. He was hired by the renowned Gilbert Lewis, Dean of the College of Chemistry at Berkeley, to be his laboratory associate from 1937 to 1939. He was appointed Instructor in Chemistry at Berkeley, and moved up the academic ladder to be Assistant Professor, Associate Professor, Professor, and Chancellor. Seaborg temporarily left Berkeley in 1942 to take a high position in the Manhattan Project, which developed the nuclear bomb. Later he received the Nobel Prize in Chemistry as co-discoverer of Plutonium. For the benefit of this chapter, Seaborg supplied copies from his diary for every page in which he refers to Samuel Ruben.

In Sam's letter of April, 29, 1935, he jokingly says, "There wuz only 2 of us there—me and Lizzie. Oh—you want to know who Lizzie is? That's my new Geiger counter." In Seaborg's diary, there are several references to important uses of "Ruben's counter." As an undergraduate student, Sam designed and built a cylindrical thin walled Geiger counter. When wrapped with a piece of blotting paper soaked in a solution containing a radioactive substance, Sam's Lizzie was an extremely sensitive device for measuring radioactivity. The Berkeley Radiation Laboratory built many copies of the counter, and Seaborg's diary shows that they were widely used for several years.

Sam became a graduate student in chemistry in January 1936. His research director was Professor Willard Libby, who later won the Nobel Prize in Chemistry for the discovery and development of carbon-14 in dating old objects. Sam carried out his work with beams generated by the cyclotrons, with neutrons, and with artificial radioactive substances made by these beams.

(a)

Sam and Dana

(b)

Sam and George

(c)

Helena, George, Connie, and Dana. Fall 1943

Photographs 3.4. Supplied by Dr. George Ruben.

CORRECTION OF A MISTAKE MADE BY E. O. LAWRENCE‡

James Cork, a visiting professor of physics working with Ernest Lawrence in 1937, passed a beam of high energy deuterons through a stack of thin platinum foils and found some unexpected and thus interesting results. When Lawrence proudly showed these results to the great physicist Niels Bohr during his visit to Berkeley, Bohr remarked that he did not believe them. Bohr's knowledge and intelligence were so high that Lawrence did not argue with him, but later asked the young physicist, Ed McMillan to review the problem. McMillan saw that the experiment involved both physics and chemistry. He selected Martin Kamen, a recently arrived visitor, to be the physicist to redo the experiments, and he asked Dean Wendell Latimer of the College of Chemistry to help him find a trusted chemist to work on the problem. Latimer recommended the graduate student Sam Ruben, and so the highly productive pair of Kamen and Ruben came together.

Kamen wrote: "We decided to start over, rejecting all the results previously obtained ... The chemistry required was extremely difficult and time-consuming, involving dissolving the (platinum) foils in boiling aqua regia, followed by an exhaustive series of repetitive precipitation's ... A complete analysis took eighteen hours of continuous effort."

After a lengthy investigation, Kamen and Ruben found that the previous results were spurious, the result of trace amounts of contaminants, which gave strong signals when exposed to the neutrons, which were a by-product of the beam. Lawrence accepted and appreciated the work Kamen and Ruben had done, and invited Kamen to joint the staff as a Research Fellow of the Radiation Laboratory at a salary of one hundred dollars per month, which Martin was delighted to accept. His assigned job was to use the cyclotrons to make radioactive substances to be used as tracers for experiments at Berkeley and elsewhere. Lawrence was particularly interested in producing tracers for medical research, which he saw as a source of funds with which to build the next large cyclotron. Also, Martin Kamen and Sam Ruben discovered carbon-14 and initiated revolutionary research on photosynthesis.

---

‡Ch. 3, p. 19. Martin D. Kamen, *Radiant Science, Dark Politics, A Memoir of the Nuclear Age*, University of California Press, Berkeley and Los Angeles, California, 1985.

SAM'S PH.D. THESIS

In May 1938, Sam received the degree of Doctor of Philosophy in Chemistry. The title of his doctoral dissertation is "Studies in Artificial Radioactivity." It has three sections: (i) One involves fundamental physics of iodine and neutrons. (ii) The second section concerned chemical reaction rates using radioactive iron. (iii) The third and longest section used radioactive tracers to study a fundamental problem in biology. It involved stomach-feeding radioactive phosphorus to rats, and after an elapsed period of time he killed the rats to determine which organs of the body had taken up the radioactivity. Ruben published six articles from his thesis research.

In May 1935, while Sam was still an undergraduate student, Willard Libby, Instructor in Chemistry, took Sam aside and advised him, in Sam's words, "to study theoretical physics and advanced theoretical math because he feels I can do it and therefore profit tremendously." (Sam's letter of May 3, 1935). Libby recommended that Sam go into the most fundamental and difficult branch of physics, a challenge Sam did not accept, but rather, he took on a comparable challenge in another field. He brought his vigor and his knowledge of chemistry and physics into the study of fundamental biological processes, making use of the new radioactive tracers. Sam was an early "molecular biologist," in that he applied the most advanced experimental methods in physics and chemistry to fundamental biological problems.

PHOTOSYNTHESIS

In and after the 1930s, many great biologists were studying photosynthesis, the process whereby plants and sunlight convert carbon dioxide and water into sugar and oxygen. Sam Ruben, Martin Kamen, and their associates were the first to use radioactive carbon in the study of photosynthesis. In March 1939, S. Ruben, W.Z. Hassid, and M. Kamen published a revolutionary series of articles, "Radioactive Carbon in the Study of Photosynthesis," in the *Journal of the American Chemical Society* **61**, 661, (1939). Further work on this subject was published in Science in 1939 and in the *Proceedings of the National Academy of Sciences* in 1940. Between 1939 and 1943, with various other co-authors, Ruben published nine articles on using radioactive tracers to study photosynthesis.

Figures 3.3 B, C & D shows Martin Kamen and Sam Ruben at work in the "Rat House," which was adjacent to the chemistry building on the Berkeley campus and adjacent to the cyclotron. Oil from vacuum pumps and grease from the machinery that operated the cyclotron stained their white laboratory coats. Martin, the master of the cyclotron, was operating the 70 centimeter (27.5 inches) cyclotron, and Sam was intently studying plant leaves that had been exposed to radioactive carbon dioxide.

The radioactive carbon used in these studies was carbon-11, abbreviated as ^{11}C, which was prepared by bombarding boric oxide in the 70 centimeter cyclotron. Carbon-11 has a half life of twenty minutes, which means that one half of it remains after twenty minutes, one quarter remains after forty minutes, one eighth remains after one hour, one part in sixty four remains after two hours. An experiment using carbon-11 needs to be completed within a small number of hours. A book by Martin D. Kamen‡ [*Radiant Science, Dark Politics, A Memoir of the Nuclear Age*, University of California Press, Berkeley and Los Angeles, California, 1985] gives this account of working in the laboratory with the fast decaying carbon-11.

> At the Rat House, Sam and Zev [W.Z. Hassid] would be waiting for me like sprinters at the starting gate. Beakers would be filled with boiling water or other solvents and pipettes ready to suck up measured volumes of radioactive solutions onto absorbent blotters, which would be held by tongs over hot plates and dried. All the necessary reagents and apparatus would be in place. The counter would be ticking away establishing the background activity. Each experiment had to be planned ahead in every detail so that no time was lost in confusion or delay deciding what procedure to follow. Anyone looking in on the Rat House when an experiment was in progress would have the impression of three madmen hopping about in an insane asylum.
>
> Radioactive carbon dioxide, $^{11}CO_2$, was rushed from the cyclotron to the adjacent building, called the Rat House. A portion of it was added to a large glass bell jar with plants growing inside, electric light bulbs illuminated the plants for ten minutes, and the light was turned off and the carbon dioxide flushed away. In some cases a leaf was placed in the dark adjacent to photographic film, and the leaf was so 'hot' with radioactivity that it imprinted a picture on the film showing where the $^{11}CO_2$ was in the leaf. Typically, leaves would be ground up, and portions treated with different reagents to give information about the chemical nature of the radioactive carbon in the leaf. Many experiments were carried out testing different features, with and without the light, with leafy plants, or with green algae.

Sam and his team developed a new experimental method to trace the course of carbon in photosynthesis, and from the first they made measurements that overthrew established ideas. Based on references from 1943 or earlier, a college textbook of biochemistry¶ stated:

> The primary photochemical reaction, involving the absorption of light by the pigments, is usually represented by a formulation we owe to Baeyer:
>
> $$CO_2 + H_2O \rightarrow H_2CO + O_2$$
>
> Carbon dioxide plus water yield formaldehyde and oxygen.
>
> $$6\ H_2CO \rightarrow C_6H_{12}O_6$$
>
> Six formaldehydes combine to form one glucose, a sugar.

I have heard it said that this theory is so simple and elegant that it should be true, but Sam and associates found that virtually no formaldehyde was formed, overthrowing this long-standing theory. Also, Ruben and associates used radioactive hydrogen and heavy oxygen (^{18}O) as tracers to study the mechanism of photosynthesis. They found that oxygen produced by photosynthesis comes from water, not from carbon dioxide as many others had said. Knowledge of this work quickly spread over the world of biology.

LIFE Magazine gave an account of this research in its issue of October 20, 1941, including several pictures of Ruben and Kamen's laboratory. The *California Monthly* of January 1943 featured an exciting article concerning this work.

CARBON-14

Although a large amount of new information was obtained using the short-lived carbon-11, there are processes in biology that occur with a time scale much longer than two hours. Physicists predicted the existence of a heavier form of carbon, carbon-14, which should have a half life of several years, but none had been able to find it. Since 1935, Martin Kamen had tried repeatedly to make carbon-14.

Martin Kamen and Sam Ruben produced carbon-14 by heavy, long-term pounding of graphite (pure carbon) with a beam of particles called deuterons in the cyclotron.

Glen Seaborg wrote in his diary of Wednesday, February 28, 1940. "At the evening Nuclear Seminar, I learned that Sam Ruben and Martin Kamen

¶Ch. 3, p. 16, *Textbook of Biochemistry*, Benjamin Harrow, W.B. Saunders Company, Philadelphia and London, 1946, page 205.

apparently have discovered a long-lived (half-life of years) carbon isotope by intense bombardment of carbon." [The Seaborg diary on page 576]:

In later work, they found that carbon-14 had half-life of more than a thousand years. With this much knowledge of its properties, it appeared that carbon-14 could be made from addition of slow neutrons to ordinary nitrogen-14. Kamen and Ruben placed large carboys of dilute solutions of ammonium nitrate near the 152 centimeter (60 inch) cyclotron where there was a large number of neutrons emitted by other uses of the beam. When nuclear piles were developed, large amounts of carbon-14 were made, and its half-time was found to be more than five thousand years.

Sam Ruben's Contribution to Chemical Warfare Research

Beginning in early 1943, Sam ceased applying top level chemistry and physics to fundamental problems of biology; he applied top level chemistry and physics to practical chemical warfare research. He and Ken Pitzer became head of a National Defense Research Committee (NDRC) project at the University of California, which paralleled the work being conducted by Professors Roscoe Dickinson and Don Yost at Caltech. The project involved laboratory studies, development of instruments, and studying the travel and dissipation of gas clouds in outdoor areas. The group under Sam's direction built novel instruments for measuring wind speed and the concentration of gases in air.

The Chemical Warfare Service had a large installation at Dugway Proving Ground in the salt desert near Tooele, Utah, and it invited University workers on chemical warfare projects administered by NDRC to visit Dugway to see their facilities and to discuss their problems. Sam Ruben visited Dugway in the spring of 1943 and heard a discussion of the anomalous toxicity of phosgene. It was thought that phosgene, Cl_2CO, reacted with water in the lungs to form two units of hydrochloric acid, HCl, but animal tests showed phosgene to be much more lethal than two units of hydrochloric acid. There were various guesses: Did phosgene break through the lung barrier to enter the blood stream? Did phosgene react with lung tissue in a way more complex than its reaction with water? Sam proposed to make radioactive phosgene from carbon-eleven, expose rats to it, and see where in the body the carbon went. This program combined Sam's work as a graduate student when he studied how radioactive phosphorus spread through the bodies of rats and his work with carbon-11 in his photochemical research.

ANDY BENSON

Dr. Andrew Benson (Figure 2.1) was an Instructor in Chemistry at the University of California during 1942 and 1943, and he was a member of Ruben and Kamen's team for their pure research but was not a part of the war work. Quotation from a letter written by Andy Benson to me, March 3, 2001:

> I received a letter from Department Chairman Joel Hildebrand in Berkeley inviting me to join the faculty of the Department of Chemistry. The Salary would be $2,000.00 per annum. That was nearly $167 dollars a month! It was a stellar opportunity… Linus Pauling and Wendell Latimer, Dean of the College of Chemistry, had contrived to provide support in organic chemistry for a young and rising star in the Berkeley faculty, Chemistry Instructor Sam Ruben, who was on the verge of being advanced to Assistant Professor. So, when I arrived, Latimer directed me to the Rat House where Sam welcomed me to a small Office/Laboratory and mass of dirty glassware freshly deserted by Henry Taube (Nobel Laureate, 1983). He introduced me to the Warburg Apparatus in the dingy lab upstairs lighted by bare clear light bulbs. He handed me his copy of *Manometric Methods* by Burris, Stauffer and Umbreit and explained that he needed to know more about the ratios of photosynthesis to respiration in his Chlorella.

Andy discussed the working hypothesis for the anomalous toxicity of phosgene and helped to carry out the experiments:

> Phosgene, a doubly noxious carbonic acid chloride (Cl-CO-Cl), could react with two proteins, I reasoned, and the novel cross-linked or double protein could elicit an immune response in the lung and consequent accumulation of fluid. So I proceeded, with Sam's encouragement, to synthesize phosgene from the carbon eleven ($^{11}CO_2$) produced by Martin Kamen in the 27.5 inch (70 cm) cyclotron near the Rat House, where I and Sam and Bill Libby had our laboratories. Mine, of course, amounted to less than eight feet [2.4 m] of bench in my office, which I had inherited from Henry Taube. Reduction of carbon dioxide over hot zinc powder was easy and quick. Then, addition of chlorine yielded the phosgene, all in about twenty minutes. We administered this to a poor rat and began to determine if the radioactivity was protein-bound as I expected. [A.A. Benson correspondence.]

* * *

Kenneth Pitzer received his Ph.D. at Caltech, was appointed as Instructor at Berkeley in 1935, and, over time, was promoted to become Professor, third Dean of the College of Chemistry, President of Rice University, and President of Stanford University. In a letter to me, Pitzer explained the history of the Rat House and how it came to be named:

> The building was built for G.N. Lewis when he first came to Berkeley and recruited Gibson, Latimer, Randall, etc. They used the building until Gilman Hall was finished. It was after the chemists moved into Gilman Hall that the old temporary building was used by some other, biology or agriculture, program to house a colony of rats. Thus the Rat House name came from the early 1920's and did not involve chemists directly.
>
> When I first came to Berkeley in 1935, chemistry had re-occupied the building. Sam Ruben and Bill Libby had space there and probably some others.

* * *

Quotations from S. Ruben, A.A. Benson, C.N. Rice, and T.N. Norris, "The Physiological Action of Phosgene." Unpublished report written by T.N. Norris, October 22, 1943. Chemistry Department, University of California.

> Research on the physiological effects of phosgene has yielded only an incomplete understanding of the reason for the extreme toxicity of the substance. It has been generally supposed that phosgene because of its ready solubility in organic materials, was able to penetrate the cell walls in the lung tissue, there to undergo hydrolysis to CO_2 and HCl, and that the HCl so formed was the cause of the severe edema observed. Death has been attributed to the edema, which fills the lungs with fluid thereby interfering drastically with oxygen exchange.
>
> Though (edema) would be capable of causing death in many cases it is not . . . entirely satisfactory as a general explanation of the extremely poisonous nature of the gas . . . With a view to testing this hypothesis, experiments were undertaken with phosgene tagged with radioactive carbon. Evidence was found that, in the case of rats, a large fraction of the gas inhaled failed to undergo hydrolysis and apparently reacts with proteinaceous material.

Several pages later:

> To summarize, it appears that only about two-thirds of phosgene inhaled by a rat undergoes hydrolysis, that most of the other third reacts with material in the blood, and that at least 50% of this third reacts with

Figure 3.5. John R.Thomas, 1944. Picture by Army photographer.

> plasma protein ... Experiments are projected to test this hypothesis and to characterize better the nature of the substance or substances formed in the reaction of phosgene with organic matter of the organism. Such knowledge could make possible the deriving of a more fruitful method of medical treatment for cases of phosgene poisoning than has been available hitherto.

FIELD TESTS BY NDRC GROUP AT BERKELEY

In addition to laboratory work on phosgene, Sam and his group designed and built instruments for analyzing poison gases out-of-doors and for measuring air temperature and wind speed. John Thomas was a member of Sam's group. As a contribution to this chapter, John Thomas described the field work Ruben did on the Berkeley Chemical Warfare project, and he recounted some amusing stories.

WRITTEN BY JOHN THOMAS IN 1997 FOR THIS BOOK

Following our discussion the other evening I tried to organize my memory about our NDRC work during the war at Berkeley and Caltech. Not an easy task, but as I remember, the following people were involved:

The Berkeley project was headed by Sam Ruben and Ken Pitzer. Ken left shortly after the project started to head up a special project in Washington D.C.The two first lieutenants were Bill Gwinn and Tom Norris. The "grunts" were [graduate students] John Huston, Kent Harmon, Stan Winkleman, Clive Countryman, me, and a fellow whose name I cannot remember.

I entered Berkeley in January 1940 as a forestry major. During the summer I worked on blister rust control and that fall had Seaborg as a Chem 1B lab instructor that semester. The combination convinced me that, as much as I enjoyed the mountains, I would probably enjoy Seaborg's type of chemistry even more. So I switched.

I took courses over the next several summers and several by examination so I was able to graduate after the spring semester of 1943. As a senior I signed up for a senior research project with Sam Ruben. l cannot remember how I got to know him, but I knew that I wanted that experience. I started work on the NDRC project while finishing up my course work.

It was at that time that I became acquainted with Bill Gwinn, Tom Norris, and Ken Pitzer, although Ken had been my freshman Chem 1A lecturer. John Huston and Kent Harmon joined the project about that time. Bill was an instructor, and I believe Tom was too, although it is possible that Tom had just finished his thesis work with Ruben. Everything I knew about rats, C-11, C-14, and photosynthesis came from being around these people. I was not involved with any of the studies. I recall that Martin Kamen was around also, but, as he was not involved in the war work, I had little to do with him.

Sam Ruben was a very charismatic man, and I got along well with him. Probably because I was truly full time, with no other responsibilities of any kind, he called on me for "gofer" work at all hours, and, like so many people, I admired him greatly.

As you know Ken was called to Washington almost immediately and was gone by the time we did our first experiments. This took place in the Yolo Bypass which I believe Ken had made arrangements for. We camped at the site in order that we be available when the conditions were best. This came primarily in the wee hours of the night, although we ran some tests at less favorable times to prove the point. We used butane to simulate phosgene and released the gas in a line source which helped simplify the interpretation. We also generated titanium dioxide fogs and observed the cloud travel. It was fun, but I never knew whether it served any real purpose or not. We worked up data and set up new experiments during the day and usually went into Davis to have dinner. Then we would play cards in the one lighted tent we had for hours until the inversions got strong enough to warrant an experiment.

After about three weeks we returned to Berkeley, reworked our instruments, and Ruben completed arrangements with a Mr. Buck who I believe was on the faculty of the forestry department. In any event he was in charge of the University-Forest Service experiment station out of McCloud on the slopes of Mt. Shasta. They were studying micro-meteorology as it influenced forest fire suppression. Most of the group then left to go to Mt Shasta via Highway 1. I was left at Berkeley to work on the instruments. Their purpose in going up Highway 1 was to scout out a suitable beach for doing experiments. Maybe at this time there was thought about using phosgene, but I doubt it. In any event they had a miserable trip. It took far longer than expected due to the curvy road. I remember something like 13 hours, and they did not find anything better than nearby Stinson Beach.

You know about Mt. Shasta, a delightful place. Both Bill's wife and Tom's wife were with us most of the time, so we ate pretty well. Before leaving Berkeley, Ruben had given me the job of going to the rationing board to get additional food stamps for meat, butter etc. I, of course wouldn't tell them what we were doing, and I never understood why they did not tell us to use what (that is, ration coupons) we already had individually. In any event I must have made it sound like the war depended on our getting extra stamps, because they gave us enough to feed an army. Latimer told me later that one of the board members was a neighbor of his, and they got a laugh from my performance. [At the time, John was 21 years old]. The grocery store in McCloud, a logging town, had a superb bakery and English-like cream that you served by the spoonful. Really great!

Tom Norris loved his coffee and kept the pot going continuously. He rarely threw out the old grounds, just added new. So, instead of adding coffee, we started adding pine needles. (We never drank much coffee and gave it up during this period). After several days Tom started commenting about the coffee and saying that perhaps he should start a new batch and maybe clean out the pot. However, it was several days before he did so, at which time the grounds were largely pine needles. At that time we confessed.

Pitzer commented about the extremely strong inversions we measured at Shasta. You probably experienced them when you visited. They gave rise to two items of interest. One was the mercaptan release, and the other was the substantial cloud of butane that we could build up. My memory is that we were still using the line source. In any event Bill and Tom started worrying about the possibility of igniting the cloud with a spark from the mercury cup anemometer. It was decided that the risk was low, but I think we all still worried. Bill's son asked me if it was true that we blew up five gallons of mercaptan with a stick of dynamite.

I told him that we released five gallons of mercaptan, but I did not tell him that Tom Norris poured it out...

When we arrived at Mt. Shasta, Buck's number one assistant was Clive Countryman, an expert in micro-meteorology and the associated measurements. He was a very quiet fellow but quite smart and 100% devoted to business. He stuttered rather badly which added to his quiet demeanor. Ruben, recognizing his worth, hired him immediately. He stayed with us until after Panama. I cannot find him in the phone book but will see if the forestry department has any record of him.

About the time we reached Mt. Shasta, Stanley Winkelman joined the project. I told you earlier that I knew he had connections with Caltech which led him to us, but I did not remember until talking to him today that he spent a good deal of time there on Yost's project. He remarked that they shot off sulfur dioxide bombs and made meteorological measurements in the desert. He mentioned several names of other workers on that project, but I do not recall running into any of them. He said most of them went to Los Alamos (nuclear bomb work) when they finished the gas work (when Yost shut down his project). Stan was very bright and willing to work and so fit in well with the group. He stayed through Florida, and deciding that a prolonged stay in the jungle was not for him, he obtained a commission in the navy and served out the war on a destroyer.

I have been worrying over where we started using bombs in our work rather than a line source. I know we did not use them in the Yolo Bypass, and I am sure we used them exclusively at Stinson Beach. Stan is certain we used them also at Mt. Shasta. I believe we did, but it is also my recollection that we used the line source at times. Stan reminds me that we illegally carried a box of blasting caps through the Caldecott tunnel. I also have a distinct memory of carrying a case of dynamite across the Bay and Golden Gate Bridges, which was a no-no too!

The mystic, whose name I cannot remember, became a problem at Shasta. We slept outside in sleeping bags, and it was not uncommon to wake up and find this guy sitting in the middle of his sleeping bag with arms and legs crossed, perfectly quiet. One night I got up and held a flashlight about an inch [2.5 cm] from his open eye for several minutes. He did not move or blink. He was really out of there. Because we had no restroom facilities, and since Bill's wife and Tom's wife were with us, one side of the camp was reserved for men, the other for the ladies. Unfortunately, the mystic was given to building shrines in the woods wherever the gods dictated. Consequently the ladies were frequently startled in the middle of the night to run across this guy in a trance on their side of the woods. I think this is what finally caused Bill to terminate his appointment after we returned to Berkeley.

In October we finished our work at Shasta, and while Ruben and Harmon returned to Berkeley, the rest of us prepared to move to Stinson Beach. Ruben had fallen asleep at the wheel on the way home [from Mt. Shasta, when I was there] and had broken his arm [George Ruben said it was his wrist]. However he was in a great hurry to test our instruments on phosgene. I always thought this was in anticipation of going to Panama, but it could have been in anticipation of going to Bushnell, Florida. In both places we were to use real war gases for the first time outside of several brief excursions to Dugway.

My recollection is that Gwinn, Norris, Winkleman, Countryman, and I went on to Stinson Beach. During that period the beach was closed to the public because it was a potential landing site for enemy people. At night it was patrolled by Gl's with attack dogs so we had to arrange night time experiments carefully with the local military. They were really quite helpful, however, providing us with four-wheel drive trucks to get our equipment out on the beach. Even so we continually got stuck in the sand and it took a lot of effort to do things. I believe it was here that an army captain, ignoring our experience, drove his jeep out near the water line at low tide and got stuck. We watched him struggle as the tide came in eventually covering the jeep in its entirety. He was not one of our favorite people. The wind and climate made day time use of gas ineffective but, as I recall, some nights we got relatively good results. It would be fun to see what the reports had to say. As I mentioned earlier I never saw one of our reports from anyplace.

RATS

To do the laboratory work on rats, Sam needed some phosgene. Although not on the NDRC project, Andy Benson, working in Sam's laboratory in the Rat House, provided phosgene for the rat experiments, as he described it:

> In our experiments, Sam and I had need for small amounts of phosgene. No steel Lecture Bottles of laboratory gases were available then, only a sawdust-filled wooden box of ancient Kahlbaum five hundred ml ampoule bottles of liquid phosgene, which had lain, since the twenties, behind the Old Chemistry Building where they could be quite cool. Phosgene boils at 8°C and could be managed at ice temperatures well enough. But, we had no fume hoods. The situation was not very healthful … I constructed some small (100 ml) metal pipe bombs with valves for the phosgene and transferred it carefully from the ampoule, cooled in ice water. I did this several times for Sam. [A.A. Benson correspondence to me.]

Peter Yankwich, a graduate student on Sam's project, recalls that these ampoules of phosgene were much smaller than 500 ml:

> My recollection of the Kahlbaum ampoule is pretty clear. The body was about 10 cm long (4 inches) and 5 cm in diameter. It had a rounded bottom and tapered quickly at the top into a tubular neck about 8 cm long and perhaps 1 cm in diameter. The ampoules were filled up to the beginning of the taper-down portion. I believe the volume of liquid phosgene contained was about 125 ml." [Personal correspondence.]

On Sept. 19, 1943, Sam sent a letter to Andy Benson mentioning several aspects of their activities.

Dr. A.A. Benson Sept. 19, 1943
Civilian Public Service, Camp No. 37
Coleville, California

Dear Andy,

Congratulations on your youngster!

Wagner was here yesterday and told me about him.

I am very sorry to have been so long in writing. I have been out of town for the most part and my arrivals and departures have been too close to allow me to do the things I would like.

Our work in Shasta has been going very well and I feel that we have accomplished something worthwhile. The work was dirty and not easy. I can tell you that I'm glad it's over with—at least that phase.

The Washington Selective Service Office turned down Latimer's request for your transfer—which is bad.

I certainly wish it could arranged. Perhaps you start things going from your end.

Rice and Yankwich have been making some progress on the problem. CNCl gives somewhat similar picture with an appreciable amount in protein form, also precipitated by $(NH_{4)})_2SO_4$.

With C-14 we hope to isolate this fraction and purify it.

There is nothing happening on photosynthesis around here or any place else as far as I can gather.

While I hope I can teach next semester, it looks as if I will not be able because of NDRC.

Sincerely,
Sam

PS. I hope I will hear from you as often as you can find time to write.

ATTITUDES TOWARDS LABORATORY SAFETY

The expression, "gung-ho", implies reckless optimism, with a feeling of superiority and self importance, "Out of my way, here I come, ready or not." Ernest Lawrence's invention, the cyclotron, was enormously successful. "All the world was beating a path to his door." In the physics and chemistry departments at Berkeley, many professors, professional research workers, administrative officers, and especially Sam Ruben were swept along with the stream of exciting new fundamental data and were "gung-ho" to a large degree.

During 1942 and 1943 I noted radically different attitudes towards laboratory safety at Berkeley and at Caltech. The Rat House was an old wooden two-story building, adjacent to the building with the 37.5 inch (95 cm) cyclotron and adjacent to the chemistry building. The main purpose of this cyclotron was to produce new radioactive materials, which the research workers analyzed and characterized. "It had no fume hoods!" As Andy Benson said with sarcastic understatement, "The situation was not very healthful." The story is told: One day, Sam wanted to calibrate his Geiger counter but found it gave an extremely high counting rate. Sam caught on, took off his laboratory coat, and threw it in a corner of the room. The counting rate went down, but it remained much higher than normal. Sam took of his shirt and, later, pants and underwear, and he threw them in the corner. The rate went down to its usual value. Safety considerations were so slack that Sam did not know that for an indefinite period his clothes carried hot radioactivity.

The freshman chemistry laboratories at Berkeley had no fume hoods, but at the far end of each laboratory there was a small open-air porch, which students crowded into to do the most hazardous parts of their experiments. The freshman laboratory used hydrogen sulfide gas in large amounts, and hydrogen sulfide is more poisonous than hydrogen cyanide. As far as safety was concerned, the attitude at Berkeley appeared to be that research workers and students should take care of themselves.

Caltech had high standards of safety and protection of research workers and students. The new Crellin Chemical Laboratory, finished in 1939, had at least one large fume hood in every laboratory. There were small downdraft fume hoods along the center of each work bench in the freshman chemistry laboratory, so that every two students had their own mini-hood. Before carrying out war-gas research, Professor Roscoe Dickinson asked for doubling the pumping capacity for two fume hoods in 65 Crellin and raising the height of the fume exhaust on the roof. Pauling found the money, and the jobs were swiftly completed. Dickinson issued "safety sheets" for handling individual poison gases and for use of the various instruments in the laboratories; and in 121 Gates he provided two open pans of a salve to be applied to the skin in case of any contact with hydrogen fluoride. He required each laboratory worker to carry a gas mask on his belt when working with poisons.

One Day, Sam

On September 27, 1943, Sam Ruben, Kent Harmon, and Robert Vogel came into their Berkeley laboratory in the Rat House. They planned to calibrate their new hot-wire gas-meter against phosgene. The rest of Sam's crew was busy dismantling the camp at Mount Shasta. Their next project was to help an Army group carry out tests using real phosgene bombs at Stinson Beach, California, and those field tests needed a method to measure phosgene gas concentration in the atmosphere.

Andy Benson had been drafted as a conscientious objector and was fighting forest fires in northern Nevada. Sam was disappointed to find that Andy's metal tank built to contain phosgene was empty. Sam had a handicap that day. He had broken his left wrist in an automobile accident during his return from the project on Mount Shasta. This was the day Malcolm Dole and I had just missed meeting Sam at their work station on Mount Shasta (Compare Chapter 2). Andy's method required cooling the source glass ampoule in ice water, scratching the side of the narrow glass tube close to its upper end, fitting a rubber tube over the glass tube past its scratch mark, clamping the rubber tube tightly with wire, connecting the rubber tube to a special expansion bulb, crushing the glass tube at the scratch mark, and passing the liquid phosgene through an open metal valve into the ice cold metal tank. The operation was a complex and dangerous one.

Kent went to the sawdust-filled wooden bin behind the Old Chemistry Building and brought back one of the old glass ampoules containing liquid phosgene. The three proceeded with the planned series of operations, and they laid out the equipment they would need for the job. First Sam slowly, carefully dipped one end of the phosgene ampoule into a Dewar flask of liquid air in order to lower its vapor pressure and make it safe to transfer. Instantly, the soft glass container cracked and split open. Still at room temperature and under pressure, liquid phosgene squirted into the liquid air. The leading slug froze solid, but in doing so, it boiled off a large amount of liquid air. This sprayed liquid phosgene high into the air which fell back as a dense cloud of cold air and vaporized phosgene. Sam, Kent, and Robert got out of the room as fast as they could and rushed away from the building. As Andy Benson said, "They lay down on the grass by Strawberry Creek."

From Peter Yankwich to Andrew Benson, e-mail 13 November 1996:

> My recollection of that day is still very vivid. John Huston and I shared a lab near the one in which Kent Harmon and Sam were working. I don't know where John was that day, but Kent was in the room with Sam, got a serious whiff of phosgene, and was hospitalized for several days. I was in what used to be Eshleman Hall that day, courting my future wife, who was an Assistant Editor of the Daily Cal. She remembers that I was chatting away with her and her colleagues when I suddenly got a strange look on my face and bolted for the door (running to the Rat House). The wind had so shifted that traces of the phosgene cloud, blown at first toward the Greek Theater, were coming down Faculty Glade, around the Student Union Building and into Eshelman Court.
>
> When I got to the Rat House, people were standing around outside; Sam and Kent had already been taken to Cowell Hospital.
>
> I did that transfer several times over the next few months, always waiting for an afternoon when the wind was steady from the West. I wore a gas mask and gloves and took all the materials across College Way into the eucalyptus grove in front of the Greek Theater. As I recall, I cooled the ampoule in ice-water, threw that away, then dropped Dry Ice into the Dewar to get the vapor pressure really low. When the phosgene was really cold, I made a circumferential scratch in the tube of the ampoule; pushed the tube carefully into a rubber hose connected to the pipe bomb you had made (it had been evacuated); lay the assembly on the sloping ground so that the pipe bomb was downhill in the cold Dewar while the ampoule was in the (warmer) air uphill — then, holding my breath, opened the valve on the bomb and crushed the ampoule tube through the rubber hose using pliers. Transfer of the liquid was fast and complete. When the transfer appeared to be over, I closed the valve, pulled the ampoule out of the hose, thanked God for my survival, and returned to the Rat House. I purified the phosgene by doing a pair of vacuum transfers on my vacuum line.
>
> In later years, I thought about those transfers and wondered why I chose to save a little time doing them that way instead of working out a really safe procedure. I guess the no-hoods reason is the most compelling.

Ambulances had come quickly, and carried the three to the nearby university hospital. Robert Vogel had been close to the open door when the accident occurred, and he received only moderate harm. Kent Harmon inhaled enough phosgene gas that he developed serious lung damage within a few hours, but he recovered after several days in the hospital. Sam had taken a heavy dose, and despite heroic efforts by the doctors in the hospital, he died early the next day.

* * *

The Coroner's report stated:

> We the jury do find that the name of the deceased one was Samuel Ruben, a native of California, aged 29, and that he came to his death on Sept. 28, 1943 at Memorial Hospital, Berkeley, Alameda County, California, and we further find that the death was caused by pulmonary edema, massive and diffuse, due to inhalation of poison gas, dilation of right ventricle, suffered accidentally in his regular line of duty in the Chemical Laboratory on the U. C. Campus, Berkeley, when a glass vial of poison gas accidentally broke. We find death was accidental.

The newspapers reported:

> Laboratory Blast Fatal to U. C. Chemist. Dr. Samuel Ruben, 30, considered 'probably the most outstanding young experimental chemist in the country,' died yesterday afternoon in Berkeley from the effects of an explosion occurring the day before in the chemical laboratory of the University of California. All details of the accident are held secret because, the university authorities announced, the fatal experiment 'had a direct bearing on the conduct of the war.'
>
> Dr. Wendell Latimer, dean of the university's college of chemistry announced Dr. Ruben's death. He said that Dr. Ruben was engaged on official investigations on Government contracts with the Office of Scientific Research and Development when the explosion occurred.
>
> Dr. Ruben died at the Cowell Memorial Hospital on the university grounds.
>
> Dr. Ruben was born in San Francisco November 5, 1913. He was graduated from the Berkeley High School in December 1930, and took his Bachelor of Science degree in 1935 at the University Of California. In 1936 he held the James Monroe McDonald scholarship in chemistry at the university, was a teaching assistant there in 1936–37, and was an Abraham Rosenberg research fellow in chemistry at the university in, 1937–38, receiving his Ph.D. in 1938. His doctoral thesis was entitled "Studies in Artificial Radioactivity." At the time of his death he was an assistant professor of chemistry.
>
> Dr. Ruben was noted for his work on photosynthesis so much so that in its issue of October 17, 1941 the magazine Life ran an illustrated article describing the distinguished scientist's work.

He died on the eighth anniversary of his marriage to Miss Helena West. Besides his widow he is survived by three children, Dana West, George Collins, and Constance May. Also surviving are Sam Ruben's father and two sisters Mrs. Ida Liss, all of 3436 Clay Street.

Funeral arrangements have not been completed.

Newspaper report, a few days later:

Two University of California scientists who survived the laboratory explosion that fatally injured Dr. Samuel Ruben, famed chemist engaged on a secret war project, yesterday were reported recovering and may continue the work. They are Kent Harmon and Robert Vogel.

HELENA AND THE CHILDREN

Figure 3.4.C shows Helena with the three children a few months after Sam died.

Figure 3.6. David Templeton, born 1920, Professor Emeritus, U.C. Berkeley. B.S. Louisiana Polytechnic Institute (1941); M.A. University of Texas (1943); Manhatten Project, U.C. Berkeley (1943–45); Ph.D. University of California, Berkeley (1947); Fil.D. (h.c.) Uppsala University (1977); Guggenheim Fellow (1954, 1969); Dean of College of Chemistry, Berkeley (1970–75); Co-editor, Acta Crystallographica (1981–84); President, Amer. Cryst. Assoc. (1984); A. Lindo Patterson Award (1987). Member, ACS; Fellow, AAAS; Faculty Senior Scientist, Chemical Sciences Division, Lawrence Berkeley National Laboratory. Photograph taken in 1941 and contributed to this book by David Templeton, 2002.

David Templeton, professor of chemistry at Berkeley, was Helena's employer for thirty years. He writes:

> Helena West Ruben graduated from the University of California at Berkeley with a major in chemistry in 1935. After the death of her husband, Sam, she was left with three young children. With them she moved into the home of her mother on Vincente Avenue in Berkeley, where she lived until 1998. Mrs. West helped care for the children while Helena sought employment. Much later Helena was burdened by the care for both her elderly mother and an aunt. For a few years she worked as a teacher in a nearby public elementary school. Then she found employment at the U. C. Radiation Laboratory (Rad Lab, now known as the Lawrence Berkeley National Laboratory) as an assistant in the x-ray diffraction laboratory headed by David Templeton. For more than thirty years she was responsible for managing the supplies and equipment for photography of diffraction patterns, for instructing graduate students and visitors in their use, and for recording powder diffraction photographs of samples submitted by others for identification or for structural analysis. Notable among these several thousand samples were many compounds of lanthanide and actinide elements from Burris Cunningham and his students and new ceramic materials from Leo Brewer and his associates. She became skilled in certain methods for growing crystals, and she prepared some which were important to the success of early experiments on x-ray dispersion with synchrotron radiation by David and Lieselotte Templeton and others. As she gained experience and learned more of the theory, she also moved into structural research using single-crystal techniques and the computing methods made possible by the ever changing computer facilities of the Rad Lab. She is a co-author (in several cases the first author) of at least twenty publications reporting new crystal structures.

Sam Ruben Bibliography

S. Ruben, W.Z. Hassid, and M.D. Kamen, "Radioactive Carbon in the Study of Photosynthesis," *Journal of the American Chemical Society* **61**, 661–663, 1939.

S. Ruben, M.D. Kamen, W.Z. Hassid, and D.C. DeVault, "Photosynthesis with Radio Carbon," *Science* **90**, 570–571, 1939.

S. Ruben, M.D. Kamen, and W.Z. Hassid, "Photosynthesis with Radioactive Carbon. II. Chemical Properties of the Intermediates," *Journal of the American Chemical Society* **62**, 3443–3450, 1940.

S. Ruben, W.Z. Hassid, and M.D. Kamen, "Radioactive Nitrogen in the Study of N_2 Fixation by Non-leguminous Plants," *Science* **91**, 578–579, 1940.

S. Ruben and M.D. Kamen, "Radioactive Carbon in the Study of Respiration in Heterotrophic Systems," *Proceedings of the National Academy of Sciences* **26,** 418–422, 1940.

S.F. Carson and S. Ruben, "CO2 Assimilation by Propionic Acid Bacteria Studied by Use of radioactive Carbon," *Proceedings of the National Academy of Sciences* **26,** 422–426, 1940.

H.A. Barker, S. Ruben and M.D. Kamen, "The Reduction of Radioactive Carbon Dioxide by Methane-Producing Bacteria," *Proceedings of the National Academy of Sciences* **26,** 426–430, 1970.

H.A. Barker, S. Ruben and J.V. Beck, "Radioactive Carbon as an Indicator of Carbon Dioxide Reduction. IV. The Synthesis of Acetic Acid from Carbon Dioxide by Clostridium Acidi-Urici," *Proceedings of the National Academy of Sciences* **26,** 477–482, 1940.

R. Overstreet, S. Ruben, and W.Z. Hassid, "The Absorption of Bicarbonate Ion with Barley Plants, as Indicated by Studies with Radioactive Carbon," *Proceedings of the National Academy of Sciences* **26,** 688–695, 1940.

S.F. Carson, J.W. Foster, S. Ruben and H.A. Barker, "Radioactive Carbon as an Indicator of Carbon Dioxide Reduction. V. Studies of Propionic Acid Bacteria." *Proceedings of the National Academy of Sciences* **27,** 229–235, 1941.

S.F. Carson, S. Ruben, M.D. Kamen and J.W. Foster, "Radioactive Carbon as an Indicator of Carbon Dioxide Reduction. VI. On the Possibility of Carbon Dioxide eduction via the Carboxylase System," *Proceedings of the National Academy of Sciences* **27,** 475–480, 1941.

J.W. Foster, S.F. Carlson, S. Ruben and M.D. Kamen, "Radioactive Carbon as an Indicator of Carbon Dioxide Reduction. VII. The Assimilation of Carbon Dioxide," *Proceedings of the National Academy of Sciences* **27,** 590–596, 1941.

C.B. van Niel, S.F. Carlson, S. Ruben, M.D. Kamen, and J.W. Foster, "Radioactive Carbon as an Indicator of Carbon Dioxide Reduction. VIII. The Role of Carbon Dioxide in Cellular Metabolism," *Proceedings of the National Academy of Sciences* **28,** 8–15, 1942.

C.B. van Niel, J.O. Thomas, S. Ruben and M.D. Kamen, "Radioactive Carbon as an Indicator of Carbon Dioxide Reduction. IX. The Assimilation of Carbon Dioxide by Protozoa," *Proceedings of the National Academy of Sciences* **28,** 157–161, 1942.

M.B. Allen and S. Ruben, "Tracer Studies with Radioactive Carbon and Hydrogen. The Synthesis and Oxidation of Fumaric Acid," *Journal of the American Chemical Society* **64**, 948–950, 1942.

P.J. Nahinsky and S. Ruben, "Tracer Studies with Radioactive Carbon: The Oxidation of Propionic Acid," *Journal of the American Chemical Society* **63**, 2275–2276, 1941.

D. Harman, T.D. Stewart, and S. Ruben, "The Synthesis of Labelled Methyl Iodide," *Journal of the American Chemical Society* **64**, 2294, 1942.

D. Harman, T.D. Stewart, and S. Ruben, "A Study of the Menschutkin reaction Using Radioactive Hydrogen as a Tracer," *Journal of the American Chemical Society* **64**, 2294–2296, 1942.

S. Ruben, M.D. Kamen, M.B. Allen, and P.J. Nahinsky, "Some Exchange Experiments with Radioactive Tracers," *Journal of the American Chemical Society* **64**, 2297–2298, 1942.

P.J. Nahinsky, C.N. Rice, S. Ruben, and M.D. Kamen, "The Synthesis and Oxidation of Several three-Carbon Acids," *Journal of the American Chemical Society* **64**, 2299, 1942.

T.H. Norris, S. Ruben, and M.B. Allen, "Tracer Experiments with Radioactive Hydrogen. Some Experiments on Photosynthesis and Chlorophyll," *Journal of the American Chemical Society* **64**, 3037–3050, 1942.

S. Ruben, M.B. Allen, and P.J. Nahinsky, "Tracer Studies with Radioactive Carbon. The Exchange between Acetic Anhydride and Sodium Acetate," *Journal of the American Chemical Society* **64**, 3050–3053, 1942.

C.B. van Niel, S. Ruben, S.F. Carson, M.D. Kamen, and J.W. Fowler, "Radioactive Carbon as an Indicator of Carbon Dioxide Utilization. VIII. The Role of Carbon Dioxide in Cellular Metabolism, *Proceedings of the National Academy of Sciences* **28**, 8–15, 1942.

C.B. van Niel, J.O. Thomas, S. Ruben, and M.D. Kamen, "Radioactive Carbon as an Indicator of Carbon Dioxide Utilization. IX. The Assimilation of Carbon Dioxide by Protozoa," *Proceedings of the National Academy of Sciences* **28**, 157–161, 1942.

S. Ruben, A.W. Frenkel, and M.D. Kamen, "Some Experiments on Chlorophyll and Photosynthesis Using Radioactive Tracers," *The Journal of Physical Chemistry* 710–714, 1942.

S. Ruben, "Photosynthesis and Phosphorylation," *Journal of the American Chemical Society* **65**, 279–282, 1943.

Copy of one page from:
SCIENCE IN WORLD WAR II,
Chemistry, A History
of the Chemistry Components of the
NATIONAL DEFENSE RESEARCH COMMITTEE
EDITED BY W.A. NOYES, JR.
Little, Brown and Company — Boston
1948

DEDICATION. This book is dedicated to the scientists, young and old, who contributed to the work described in this history. The young men, in particular, gave of their time and energy at crucial periods in their lives without seeking and receiving public acclaim and glory. They accepted orders and carried them out faithfully, often with no knowledge of ultimate objectives and often without even that satisfaction which came to older men from knowing that their work was accepted and helped to win the war. They served their country with less recognition but just as truly as did those in uniform.

Many men of the chemical divisions of OSRD gave their lives for their country. Their names belong on the Roll of Honor along with the names of those who died in battle. To these men, especially, is this volume dedicated:

DONALD E. BOYER
ROBERT S. DONE
JOHN FEHRER
JOHN HANUSIN
CHARLES R. HOOVER
ROY NELSON HUNT
JOHN LEONARD
H. G. NELLIS
CLARENCE O'BRYAN
F. K. OVITZ
SAM RUBEN
ROY C. WOEHRMAN

CHAPTER 4

WITHLACOOCHE

BUSHNELL, FLORIDA

In 1943 the United States and its allies began the difficult task of retaking islands in the southwest Pacific from the Japanese. Typically, these areas were tropical jungles, which, of course, were different from the deserts of California and Utah and from the pine forest of Idaho, where the National Defense Research Committee (NDRC) and the Chemical Warfare Service (CWS) had tested chemical weapons and studied the travel of gas clouds. A team consisting of NDRC Division 10 personnel and CWS officers explored areas in the southeastern United States for a suitable site for field tests in semi-tropical forests. They selected, "… a government-owned tract of land in Florida, known as the Withlacoochee Development Project of the Soil Conservation Service, U.S. Dept. of Agriculture, which comprised about 50,000 acres (20 thousand hectares) and which was sufficiently remote from any towns and villages to permit the release of considerable quantities of toxic agents without danger to man or animals." [W.A. Noyes, Jr. (1948), page 235]. The Chemical Warfare Service set up a camp in Bushnell, Florida entitled Dugway Proving Ground Mobile CWS Field Unit. Bushnell is approximately in the center of the state, north and south, east and west; it is almost due west of Orlando and about halfway between Orlando and the Gulf of Mexico. Field trials using nonpersistent gases, phosgene, cyanogen chloride, and hydrogen cyanide, were scheduled for November and December, 1943.

During early November 1943, Professor Roscoe Dickinson, John Otvos, Bob Mills, I, and three from Professor Yost's former group loaded thirty Egberts (Dickinson meters), including Esterline-Angus recording meters, other equipment, and our heavy personal gear on the back of a two-ton flat-bed Ford truck. We loaded the meteorological instruments on the top of the Buick station wagon, and stuffed the back compartment of the station wagon with delicate electronic gear, various equipment, books, and some personal items. Dickinson and the six of us were headed for a project, of indefinite duration, in Bushnell, Florida. Bob Mills kept a simple, amusing diary of this trip across the country in the middle of the war, which I quote here in full.

BOB MILLS' DIARY: PASADENA, CALIFORNIA, TO BUSHNELL, FLORIDA.

Tuesday morn. Nov. 9, 1943
Pasadena, Calif.
Leave C.I.T. at 0934

Team # 1	John Otvos and Phil Hayward
Team # 2	Hal Johnston and Bob Mills
Team # 3	Mike Kraus and Ted Gilman

Truck pulled out first, and as station wagon attempted to leave circle it was so heavily loaded that the rear fender caught on the curb and pulled off the part. While this was being fixed the truck was out of sight. Drove out to junction of Colorado Street and route #19 at Shell Station. Phoned Betsy to ride in station wagon out Hill and look for Otvos. Leave station at 0957 without waiting for Betsy. Meet John at Azusa, and will take the lead, John following in truck. Ate at Green Spot Cafe in Victorville. Ham loaf luncheon $.82 and $.10 (tip).

Leave Victorville at 1:00 Harold driving station wagon; Otvos the truck at 1:15. The teams change and I take over the truck. See beer truck turned over in road. Sad sight. Refueled at Barstow and played catch with John. Proceeded with much noise and vibration until 3:40 at which time the teams change again and I take my place in the back seat of station wagon. Proceed to Needles. Reached Needles at 5:35. Truck had run out of gas 5 miles back. Stop at Shell Station and get 1 gallon (3.8 liter) to take back to truck. Ate at California Coffee Shop Chicken Fried Steak 97¢. Take a dim view of the whole deal.

Crossed Colorado river at 7:06 and enter Arizona. Switch teams and Otvos takes over truck. Arrived in Kingman at 9:30 and hunted for place to sleep. Found one vacancy at Gypsy Motel. Cabin contained 2 double beds so we pitched 4 cots and sleeping bags. Got cited by Arizona patrolman for pulling reverse U-turn in front of Hotel. Incidentally Hotel was full. Paid 50¢ for my bed. Retired in my sleeping bag at 10:45. Drove 358 miles.

Nov. 10

Got up at 7:00. Dressed and ate at Peggy's cafe 40¢. Hotcakes and egg!!!! Heard two renditions of "I hang my head and cry" by Gene Autry. Got away from Kingman at 8:12. Team #1 still operating in truck. Stopped for gas at 8:55 just past Hackenburry Wash. 9:30 I take over the truck. At 12 noon (11 am) stop for gas. Get 10 gals at Ashfork cost $2.35. Stopped outside of Williams and switched teams at (12:00) 1:00 p.m.

Just before the switch the truck had conked out on a hill. Ate lunch in Flagstaff. Steak – $1.47 + $.15. Left to get gas in the city. Bought 5 cards and mailed one home.

Switched teams and Otvos drove to Holbrook where we all refueled. Had passed Painted Desert and Petrified Forest. Are trying to make Albuquerque tonight but by 5 pm we have only come half way (250 miles) so far. See lots of Indians and lazy dogs.

Reached border inspection station at 7:10. Going to Gallup for dinner and bed. Took over truck with Dr. D. 5 or 10 miles from Gallup. Arrived Gallup about 8:15 and ate at Cafe. Ham steak dinner $.66 + $.15. Rented two cottages at Jim's Motel." Harold and I are to sleep in double bed. Swell cottages—showers and modern fixtures, garage etc. Wrote cards, took shower and went to bed at 11 p.m. after Planter's Punch at El Rancho Room $1.30.

November 11, 1943

Mike woke us up at 5:45 with fiendish delight. Ate breakfast at El Rancho at 6:30, #3 $.66. Clearance lights on truck didn't work until they were jarred on when I started at 7:15. Changed shifts after 90 miles at 9:00 a.m. Reached Albuquerque at 10:30. No automobile club. Mailed card to Naomi.

Ate lunch in Santa Rosa. Liver and onions $.66 + 0.05. Gassed up and left for the Texas border at 2 p.m. (Hal stopped farther back at Moriarity for gas and candy bars $.15) John still driving truck.

Started driving truck at 2:38. Crossed the Texas state line at 4:17 and was relieved at 4:38 (of driving I mean). Had no trouble with border officials. Stopped for gas 21 miles in Texas and played ball in 3rd state so far.

Reached Amarillo and ate outside the city limits at Amos' Chicken Steak House—Steak @$.87. Proposed plan—get gas and continue to mush on. Wrote card to Jean ???? Mail later. Finally stop at 12:15 in Elk City and rent the Motor Inn Courts all to ourselves. Drove 585 miles today and tomorrow morning I take over truck. Nice cabins—price $1.60. Went to bed at 1 a.m. and slept damn fine by myself in a double bed (really pounded my ear).

11/12/43

Mike came in at 7:30 and got us up. Ate at the Cabin Cafe. Egg and toast $.60. Very fine. Bed cost $1.60. Plan to play catch after breakfast in Oklahoma. Did almost. Took over truck at 8:50 and drove non-stop to El Reno (90 miles). Switched to Mike in truck at 11:05. Drove through Oklahoma City and stopped Biltmore hotel. Find out there is a detour between OK. City and Fort Smith so we plan to swing a little north. Stop in Sprague for lunch and truck lubrication and gas. Buy cookies at bakery 15¢. Eat roast beef lunch at HiWay Cafe 60¢. Notice most of the women in these parts are particularly sad. Left Sprague at 2:42. Mike still in truck. Took over truck with Harold at 6 p.m. and crossed Arkansas border at 7 p.m. after traveling over rough roads (no safety-belts were broken). Stopped at gas station in Fort Smith, gassed up and went to "Snack Shop" for dinner $1.22 + $.10. Best steak ever. Left Fort Smith at 9:35 and headed for Russellville, arrived and refueled and switched to route #64. Stopped several miles out of town at tourist cabins. Dr. Dickinson, Mike and Hal stay at the D. Bros. Modern Cabins while the other three stay a few 100 feet (30 m) down the highway at other cabins. Retire at 11:45. Beds soft and furnishings rural varying to crude. Bed cost $1.50. Crapper standard equipment. Sniffed at several females of the opposite sex in the tourist store and office. On closer checking I find them rural too. Wonder what's wrong with rural gals. Decide nothing, but had better sleep on it—the thought and decision I mean.

11/13/43

Arose at 7:45 when Mike made a hell of a racket knocking over chairs and waste basket. Dressed etc. and went in for a swell catch with John. Weather frosty and invigorating. Mush on to find breakfast at 8:25.

Stopped at Warren's Cafe in Morriston Ark. for breakfast. Ham and Eggs. 56¢. Mailed card to Aunt Peggy. Noticed a particular southern accent of waitress. "How do you all want yor eggs?" Left at 9:26.

The fan in truck threw a blade through the cowling. Got new fan installed at Ford dealers in Morrilton cost $7.50. Bought correspondence supplies at P.O. and mailed card to Aunt Anna. Bought fuses and talked to couple Negroes about Pasadena, Buicks, etc. Left Morrilton at 11:30. Truck went ahead to determine the best speed - least vibrations. Determined at 45 mph. Stopped just east of N. Little Rock at Broadway Cafe for lunch. Ordered pork plate lunch. (John driving truck) price $.46 + $.10. Left at approx 2 p.m. Passed Bear Lake full of cypress trees. World's largest fish hatchery. See lots of cotton fields and Negro tenant houses. Weather still very clear. Stopped at Gulf station for fuel and Ted relieves himself while John and I played catch and proceeded to lose the good baseball in sewage swamp back of garage. John still driving. Took over truck at 4:10 and crossed Mississippi river at 5:35 into Memphis. Cars and trucks were spaced 150 ft. crossing the bridge so as to distribute the load. For the 10 or more miles approaching Memphis I could see a big fog and smoke bank hanging in the sky. The bridge crossing the Mississippi was very high and the surrounding land on both banks of the river was very low. Crossed 4th state boundary. Crossed into Mississippi at 6:30 (my 5th state) was relieved at 7:10 by

Mike. Getting damn hungry. Plenty of gas stations—no eating joints—yes there was Staffords Cafe, Holly Springs—Steak dinner @ $.82 + $.10 accompanied by more southern clatter and stuff—you all. Got our respective asses in gear at 8:15. Where too. Dr. Dickinson quote 'Stick up for your race, Mike.' Rolled right into Tupelo, Miss. where we gassed up and then drove to the Rex Plazza Motel. Had single room to myself. Plenty snazzy. Took shower, shaved, listened to Mike's radio and turned in 11:30.

Tomorrow Sunday 11/14/43

Was awakened by Mike at 8 a.m. Dressed and ate at Rex Cafe. #3 $.65. Pin ball machine was no damn good. Played catch in Miss. Packed and got away at 9:15 with Mike still driving. Price of room $ 1.75. Crossed into Alabama at 10:10. Mike still driving truck. Switch teams at 10:15. Started driving truck at 1 p.m. Reached Birmingham 1:25 and ate at Casino Rest. Pork @ 70¢. Left 2:30. Changed drivers at 5:20 after a hectic 100 miles; things is grim! Ate in Troy at Bus Cafe. Hamburg steak @ 70¢. Went on to Dothan by 10:30 got gas and inquired about cabins. Finally went to brick cabins, 1 mile (1.6 km) outside Dothan. Called Beautyrest Camp. Flipped with John for double bed and he lost. Slept fine under genuine horse blanket until 7:30 when Sarge Kraus disturbed the peace. Price $1.32. Dead bugs in kerosene bottle. No hot water. Played catch with John and left at 7:55. John still driving the truck. Crossed into Florida at 8:25. Stopped at Chipola Hotel coffee shop. Marianna. Ordered #7. 60¢. Reached Tallahassee 12:30 E.W.T. [Eastern War Time] Teams changed Apalachicola River. Started driving truck at 10:20 C.W.T. [Central War time]. Florida seems warmer. Sunny Fla. and hot too. Can't expand on the warmer yet. Just got here. Stopped at Perry for lunch and ate pork chop dinner at Poinsetta Restaurant 65¢ + 10¢. It was also time to switch teams so Mike took over truck and party got underway at 3:10. Fairly cloudy but warm. Meal consisted mainly of turnip greens and collard greens.

(Bob's version of the **truck fire**.) A little way out of Old Jacon, we noticed that the truck wasn't behind us. So we stopped and played catch. The truck still didn't come so we reversed our field and found the truck in a gas station with the tarp off supposedly being inspected. But it had really caught fire, and Mike stopped when some Negroes shouted at him. They doused the burning cases with bucketfuls of water (with the aid of some Negroes) and when we pulled up the fire was out and the truck was partly unloaded. The 22 shells and gun were in the midst of the fire but suffered no damage. Suitcases were badly burned and Harold lost a raincoat, fine jacket and 2 pr. shoes. Several boxes were burned along with the tarp. In the dirty ruins things were repacked. The Negroes and helpers were tipped for their aid, and the truck pulled out at 5:55 after refueling. Station wagon followed.

(I inject here my recollection of the **truck fire**:

Ted Gilman was driving the truck, I sat next to the wide-open right-hand window, and Mike Kraus sat in the middle. As we drove past a farmhouse, a man rushed out waving to us in what I regarded as a friendly gesture. I smiled at him and waved back. As we passed the next farmhouse, another person ran out and waved at us. Again I smiled and waved back. I commented to Mike and Ted that the natives here were surprisingly friendly. When next three people waved at us, Ted in his mirror noticed smoke rising from the back of the truck, and he stopped. The "friendly natives" helped us carry buckets of water to put out the fire, and later we pulled into a gas station. Ted Gilman smoked much of the time, and apparently he flipped a hot cigarette butt out the window, which landed under the tarp.)

Resumption of Bob's diary:

Crossed the Suwannee River at 6:00 and soon thereafter we were stopped at the Florida agricultural checking station. Passed and proceed on to Bushnell (@ 100 mi). At Old Town John mailed a card—'Had a hot time in the Old Town'. It was also suggested that we award Mike and Ted the silver cross for bravery while under 'fire.'

Switched drivers and John and R.G.D. took over truck. The guys in the station wagon had quite a bull session. Stopped at Williston for dinner and had a good T-bone steak meal cost $1.05 and 10¢. More bullshit and before we knew it we were in Bushnell around 10:15. Looked over hotel and since Dole (?) had things already fouled up (already!) there was no place to stay in the hotel. Dr. Dickinson took his room and Bill Thomas with six of us drove over to Brooksville to Tangerine Hotel. Hit turkey buzzard on way over. Arrived 12 midnight and John(?) and I roomed in #217. Retired 12:30. Got up. Room $1.50.

11/16/43

When awakened by blasted chickens screaming, etc. Dressed, shaved and went to Florida Cafe to eat. Notice the women in this town are much improved. Breakfast 60¢.

JOHN THOMAS' ACCOUNT OF TRIP FROM BERKELEY, CALIFORNIA TO BUSHNELL, FLORIDA:

I would guess that about the end of October or early November we finished at Stinson and returned to Berkeley to get ready to go to Bushnell, Florida. Gwinn, Winkleman, Countryman, and I made up the group from Berkeley. Before leaving Berkeley for Florida, Stan and I made a trip to Dugway. Volman and others were there and we did some field work, the nature of which I have forgotten. While working in a phosgene cloud the fludder valve fell off my gas mask and before I got it back on I had taken several breaths of gas. That evening I developed some chest pains which caused them to send me to the army hospital in Salt Lake City. The next morning I was fine and Stan and I took the train back home." (This accident occurred about two months after Sam Ruben died from phosgene poisoning.)

We met your group from Cal Tech and many of the others who later went on to Panama like Bob Brinton and Dave Volman. I am sure Francis Blacet was there too. The Chemical Warfare Corp was in charge of the operation, and I know that the Airforce contributed meteorologists because one of their wives rode back with us to Berkeley.

We drove three cars to Florida. One was a large panel truck loaded (overloaded actually) with lead storage batteries, variacs, electric motors, recorders and yards and yards of heavy copper cable. We also had a pick-up truck and a large sedan. I remember the panel truck very well. I was driving it through Alabama on a two lane road in lonely farming country with the other cars following behind. There were no other cars on the road. We were pushing to get to Florida, but I was going too fast, particularly in view of the generator. As I rounded a curve there was a farmer's dog sleeping squarely in the middle of the road. In the instant I had to decide whether to hit the dog or try to evade him, I chose the latter. This maneuver caused the generator to start swinging and on the second oscillation the trailer hitch broke which caused the truck to fishtail violently. The truck, loaded as it was, wanted to turn 180 degrees and proceed down the road backwards. It took me at least 50 yards fishtailing from the ditch on one side of the road to that on the other before I could get the truck stopped. The generator, I learned later, did a complete flip and, as Bill and the others rounded the curve they saw the generator upside down in the middle of the road and the truck in the ditch on the wrong side of the road.

Bill surveyed the damage, and the farmer, attracted by the noise, called a tow truck from the nearby small town. The tow truck got the generator uprighted and Bill went into town where they welded the hitch back together. We got the truck going again and picked up Bill, making it to Florida only several hours late. The generator never looked like much after that but it worked fine and I believe the military took it on to Panama for us.

I remember Bushnell, the swamp in which we worked, the hotel in which Stan and I initially roomed together, the private house I stayed in later, a visit by Latimer and a number of the people, but the extent of our experiments there escapes me.

RESIDENCE IN BUSHNELL

Malcolm Dole had gone from house to house the previous week to find places that would rent rooms to the NDRC group. I was assigned a room in the sheriff's house, and I moved in the day after we arrived at Bushnell. We had our meals at the local restaurant, except we got in line and ate army chow during army field tests. The army group had set up tents in a flat area just south of Bushnell, where the soldiers slept and ate. Officers lived in rented rooms, apartments, or houses in town.

Mail in and out of Bushnell went through an APO (Army Post Office) coded address. Correspondents did not know where I was. Both incoming and outgoing mail was subject to censorship.

The commanding officer of the Dugway Proving Ground Mobile Field Unit was Captain Jake Nolen, who was in his early thirties and had obtained a Ph.D. degree in chemical engineering from MIT (Massachusetts Institute of Technology). Nolen was young, technically sound, and administratively tough. Through scientific competence he maintained respect and rapport with the civilian scientists, and he seemed always on top of problems concerning military personnel. Like Woodstock during the depression, Bushnell had a number of empty stores in its downtown area in 1943. One empty store was rented by NDRC Division 10, and it provided office space and storage space for all our operations. Captain Nolen and his staff rented another empty store for their offices. Nolen's office was at the rear of the store. It was separated from the rest of the building by a partition that did not go to the ceiling, and it communicated with the rest of the room by a wide opening at the right side of the partition. Any raised voice in Captain Nolen's office could be heard in all the store.

An old-time elderly army Major, whose name I have forgotten, had a desk next to Nolen's office. He had been in the chemical warfare service since World War I. His presence made for a delicate situation. The Major outranked the Captain, who was in command of the operation. When Nolen spoke with him, Nolen said the word "sir" at least once in every sentence. Within limits, Nolen let the Major do what he wanted to do.

INSPECTION OF THE TEST AREA

On the morning after we arrived, the army had conducted a test by exploding a bomb filled with hydrogen cyanide on the ground. We were invited to go see the test area in the afternoon. Three jeeps, each driven by a soldier, took us out to the test area. About a mile (1.6 km) east of Bushnell, we turned south at Bevilles Corner onto highway 471, a rough dusty dirt and gravel road that went straight north and south. The area around Bushnell contained lakes and ponds of various sizes. We drove through Webster and Tarrytown, and then entered an area that consisted primarily of wetlands of all sorts: ponds, lakes, swamps, and the Little Withlacoochee River. Bald cypresses with drooping Spanish moss grew in shallow water. There were masses of cattails and tall swamp grasses. Highway 471 was a built-up pile of dirt that sliced straight through the wetlands, with an occasional bridge. We continued along Highway 471, passed the road that went to Bevell Place, and then turned to the east at the next cross road. After about a mile and a half (2.4 km) on this narrow winding road, we entered the great Withlacoochee Hammock. From Bushnell to the test area, the total distance by road was about nine miles (14 km). This route is given by the map, which is Figure 4.1.

By definition, a hummock is a rounded knoll or hillock. "Hammock" is a variant of "hummock" in southeastern United States. In Florida, hammocks are where the land rises above the surrounding wetlands and has deep fertile soil with hardwood vegetation, including live oaks, magnolias, liquidamber, and some pines. Palmettos flourish at the ground level, and occasional pine trees rise well above the forest canopy. The Withlacoochee Hammock is indicated in Figure 4.1 as the uniform gray area in the lower panel of the map. The hammock is two miles wide at its widest. The white zone to the east of the hammock represents the meadow that also was used in the army tests. The lines on the lower panel are roads, some of which were merely two wheel-ruts between the trees.

Figure 4.2. shows two views of the forest, one above the canopy and twenty-three meters (seventy-five feet) above the ground, and one at ground level. Army photographers took these pictures on July 5, 1944.

As we approached the test area, we had our gas masks ready to put on at the first smell of gas. Instead of encountering the residue of the war gas, we saw a dense cloud of smoke. The forest was on fire. A sergeant stopped our jeeps. He told us to get out and help fight the fire, but he excused Professor Dickinson. Professor Dickinson ordered me to stay in the jeep, but he got out and went with the rest. The fire was not an extensive one, soon it was put out, the soldiers and NDRC civilians came back to the jeeps, and we drove to a storage area in the forest, where the ground had been cleared of brush. We soon learned what started the fire. The test planned that day was exploding a hundred-pound [45 kg] bomb on the ground, the gas was hydrogen cyanide, and upon detonation the gas ignited and set the

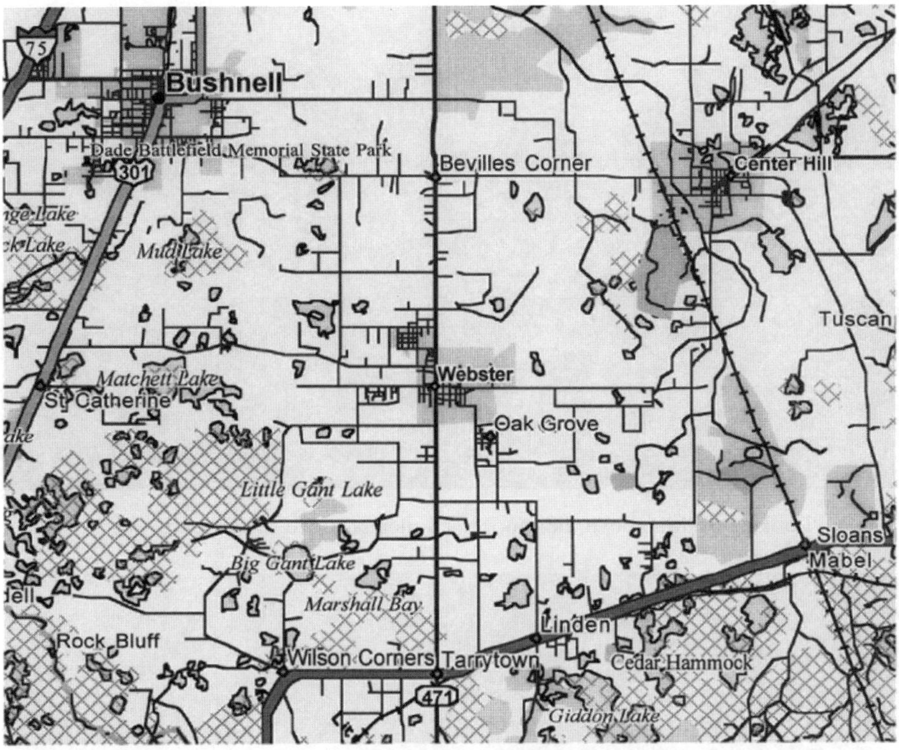

Bushnell → Bevilles Corner → Webster → Tarrytown →

Change of scale:

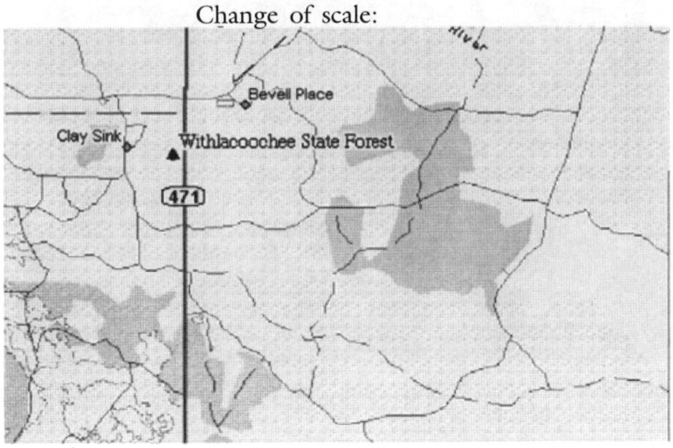

Turn east below "471"

TARGET AREAS
Dense hardwood "hammock"
Meadow east of "hammock"

Figure 4.1. Map Of Bushnell, Florida, And Target Areas.

Courtesy of DeLorme, "Street Atlas USA."
DeLorme, Two DeLorme Drive,
P.O. Box 298, Yarmouth, ME 04096 www.delorme.com, 207-865-1234.

Looking South, above the canopy, 23 meters above the ground.

At ground level, Southeast of meteorological tree tower.

Figure 4.2. Views of Withlacoochee Hammock, 1944. Photographs taken by army photographer.

woods on fire. Dickinson interviewed the officer in charge in some detail, and the officer said that a large fraction of bomb tests with hydrogen cyanide had resulted in burning up the gas. Professor Dickinson suggested a solution: add about five percent gasoline to the hydrogen cyanide. Put out a fire with gasoline? That seemed to be crazy. Later the army group tested the professor's suggestion, put five percent gasoline in the hydrogen cyanide, and it burst without catching on fire. It took me several years to figure out why gasoline put out the flame; it must be the anti-knock ingredient in the gasoline that moderated the explosion.

After the adventure with the flaming cyanide, our three jeeps proceeded to explore the meadow area, which is the white area on the map east-southeast of the hammock. The meadow was rough with shallow gullies and small hills, had a few tall pine trees, and had several areas of low growing brush. It was much rougher than the cow pasture that we had used in Pasadena.

Army Field Tests

For the first real shoot, soldiers had marked the ground with whitewash to show where the five hundred pound (227 kg) phosgene bomb would be located and where they would later place goats. We defined a grid around the bomb where the Egberts and the hot-wire devices from the University of California would be placed. We moved in our meteorological equipment and set up our Egberts with help from enlisted men. We set up one micro-meteorological station in the rough meadow and another within a grove of tall trees, which loomed high above our five-meter-high instrument. At the forest station, the army had cleared the undergrowth for about an acre (0.4 hectare). It had an area for parking trucks and jeeps, and labeled poison-gas-filled bombs lay in a line on the ground.

Malcolm Dole was in charge of all the NDRC personnel, and he proceeded to spell out in detail what we were supposed to do. For the first test he issued the following written instructions (Government document obtained from Dugway Proving Ground in 1997 through the Freedom of Information Act and not subject to copyright):

> NDRC LABORATORY, DUGWAY PROVING GROUND, TOOELE, UTAH, FLORIDA FIELD TRIALS. Meteorological Program for Shoot No. 1
>
> 1. Dickinson and Mills proceed directly to Meadow Station taking British anemometers and bivane. Set up anemometers and bivane and make temperature mast ready. Mills commence measurements and Dickinson take jeep to point on road near L12.
> 2. Kraus go to forest station, prepare temperature mast and bivane and commence measurements.

3. Johnston go to station at H3 and start direction and velocity recorders. Mark the time on the record paper. Proceed along L row to station at L12 - M12 and start recorders. Then go to road and proceed by jeep to meadow station.
   Events 1–3 should be completed not less than 15 minutes before the shoot.
4. At the meadow Station Johnston will read a velocity profile (British anemometers at 1, 2, and 3 meters; 5 min. average) every ten minutes and will make a bivane record every ten minutes. Mills will read a temperature profile every ten minutes and make psychrometer observations every half hour (the first reading being before the shoot). Mills will keep a record of sky and ground conditions at ten minute intervals.
5. At Forest Station Kraus will read a temperature profile every ten minutes and make a bivane record every ten minutes. Dickinson will keep a record of sky and ground conditions at ten minute intervals and will frequently observe wind direction and velocity at the forest station.
6. Countryman will read velocity profiles on Berkeley anemometers every five minutes. Gwinn will see that temperature gradient recorder is functioning properly.
   Each man should provide himself with data sheets and data boards beforehand. Be sure to record the data, shoot number, and time of burst on each sheet. Any visual impressions of cloud travel should also be noted on data sheets immediately after the shoot. Check watches with that of the Technical Director before the burst.
   Signed, Malcolm Dole, Director

Also, Dole sent detailed instructions to Otvos and others who operated the Egberts and the Berkeley analytical instruments. All of us understood the workings of our instruments, and many of us felt that these instructions were more detailed than necessary. Dole sent no directives to the army personnel. He was director of the NDRC group and had no authority over the soldiers or officers. The army carried out their own procedure, including setting up rings of tethered goats around the bomb.

There were many mosquitoes in the forest and meadow. In boxes nailed to trees at places throughout the test areas, the army provided large bottles of mosquito repellent. Also, there were ticks that burrowed their heads below the skin; we lightly treated them with mosquito repellent and pulled them out with a slow twisting motion so their heads would not break off and remain under the skin. Also we could reduce the number of ticks we picked up by generous application of insect repellent to the lower parts of our pants.

The first test was with a five-hundred-pound bomb of phosgene, to which a small amount of nitrogen dioxide (NO_2) had been added to give the gas cloud a visible red color. From the meadow meteorological station where I worked, I heard the explosion that ruptured the bomb, but I was so far away that I never saw or smelled phosgene. After the test, delicate instruments were loaded into an army truck, which was sealed and left overnight in the forest parking area. Goats were loaded into other trucks and driven back to Bushnell.

A senior civilian from Dugway told me that he had walked through the test area the morning after the shoot. Near the test center, he said, green leaves on the trees were withered. He came across a large dead owl and noticed many dead salamanders, fish, birds, and other small creatures.

The army and the NDRC staff soon had the testing system well organized. For six weeks, we had three to five tests every week. Some tests involved static bombs exploded on the ground, and we had several tests with bombs dropped from airplanes. Bombs containing hydrogen cyanide doped with five percent of gasoline did not catch fire for either ground burst and air drop, confirming, so far, Professor Dickinson's prediction. The gases used included phosgene, hydrogen cyanide, and cyanogen chloride. The Caltech group placed the Dickinson meters in the forest or in the meadow along a grid around static bomb test site or around the target for an air-drop bomb. After the bomb or bombs exploded, the group put on gas masks and went in to check the instruments. After a prescribed time, they turned off the recording meters, labeled and removed the chart paper that had been used, tightly wrapped the instruments with waterproof canvas, and placed them on the truck. They brought back to town any malfunctioning instrument for overnight repairs. In one experiment, an air-dropped bomb made a direct hit on one of the Egberts.

I worked only on the micro-meteorological studies. The work for other NDRC workers was not so easy, as illustrated by this write-up by Arthur Pardee:

> I think we went to Bushnell in early Fall of 1943. We joined you and others there. The objective there was to study several gases released under jungle conditions. Our job was to set out a string of bubblers, and later other apparatuses, along trails in the jungle. After the bombs went off, or were dropped from airplanes, we went in and took samples. This was somewhat scary, especially at night because one was alone with poisonous water snakes as well as the poison gas in the dark and far from anyone. Poisoned birds sometimes dropped dead from the trees.

JEEPS FOR WORK AND PLAY

The motor pool in Bushnell would lend a jeep to NDRC personnel for the day for use during a test. A jeep seated two in the front seats and two in the rear seat. Driven fast over the bumpy road, the jeep oscillated wildly; riders in the front seats were at the center of the oscillation and were little jarred, but riders in the rear seat were tossed and spanked by the bucking car. We rotated seating assignments among us civilian jeep-riders.

The motor pool would even sign out a jeep or a weapons carrier for weekend recreational use. I checked out a jeep and went to Tampa one weekend with one soldier and with one NDRC person. We visited St. Petersburg and the gulf beach, including its open air shops and tourist traps. From an artist, I bought a signed (Harold S. Etter) watercolor painting of a beach on the Gulf of Mexico with wind-blown palm trees. The picture frame was twenty-two by twenty-seven inches, and it crowded the jeep a bit as we drove back to Bushnell. On that drive, we encountered the worst fog I had ever driven in. The fog cut off my vision of the road. My passengers took turns walking with a flashlight on the road edge until we finally came to the end of the heavy fog.

FIRST CALL FOR PANAMA

After about four weeks, we all received a letter from Professor W.A. Noyes Jr., the head of NDRC Division 10, our division. The letter stated that the tests of non-persistent gases in Bushnell would terminate by the end of December, and such work would be transferred to a new station in Panama. The semi-tropical forests of Florida were not a satisfactory substitute for the jungles of southwest Pacific islands, but there were real tropical jungles in Panama. In 1944, work in Bushnell would be on persistent gases, and a new group, NDRC Division 9, would come to Florida to carry out this line of work.

All members of NDRC Division 10 were invited to volunteer for work in Panama. I volunteered to go to Panama, but Noyes, aware of my health problem, turned me down. Captain Jake Nolen and Professor Dickinson offered me the position of "Head of the meteorology department of the Dugway Proving Ground Mobile CWS Field Unit of the United States Army." What a fancy title! Captain Nolen placed conditions on this offer: I should report directly to him, and I must take a physical examination every three months from the two army doctors on duty at Bushnell.

HOW THE MILITARY MEASURED GAS CONCENTRATIONS IN THE FIELD

In general there was good cooperation between military and civilian personnel at the base. Even so, there were often interesting contrasts between the military and civilian point of view.

NDRC scientists measured the travel of the war gases by concentric circles of automatic chemical samplers. Military personnel staked out goats on a grid of their own. The goat detail was handled by the old-time chemical warfare Major. After one large test, there was a discrepancy between the results of the chemical analyses and the location of dead goats.

The chemists said that no significant amount of gas got beyond the hundred-yard (91 m) circle, but there were two goats on the two-hundred yard (182 m) circle that had died shortly after being brought back to their pens. There was a conference to decide why we had the discrepancy between methods. The chemists defended their methods and were certain that their instruments had not failed on that day. The Major suggested that the gases might have passed the hundred-yard circle in a highly dilute state and then had come back together to be concentrated enough to kill the goats on the two-hundred yard line.

I was present at this conference and naively exclaimed, "On the basis of the Second Law of Thermodynamics, I can say for sure that is impossible."

"Well, I don't know nothing about the Second Law of Ther-mo-dy-namics, but I do know a dead goat when I see one." He laughed heavily and was joined by some others.

Captain Nolen, however, probed further: "Tell me, Major, sir, how many goats were on that two hundred yard circle?"

"Eight," said the Major confidently.

"How many died, sir, after they were brought back to their pens?"

"Two," was the reply. "Two of them died during that night."

"How many goats are in the pen that have not been exposed yet, sir?" Captain Nolen asked.

"How many do we have, Sergeant?" the Major asked his assistant.

"They's only fifteen left now, sir. We're getting sort of low on goats," the sergeant said.

Captain Nolen, who signed every purchase order, including those for goats, pushed on: "Sir, how many goats have died in the last few days that were <u>not</u> exposed in the field?"

"Why, none, I don't think," the Major exclaimed. "We haven't had any of them die recently, have we Sergeant?"

"Sir, the last few days we've had a right smart of Texas fever in them there goats," the Sergeant said, nervously glancing back and forth between the Captain and the Major.

Captain Nolen pursued the point directly with the Sergeant: "How many goats left in the pen died since the test four days ago?"

"They's been six of 'em died from Texas fever these last four days," the Sergeant said.

Captain Nolen spelled out the conclusion: "Six out of fifteen died from Texas fever in four days, and they never got near the field. That's even more than two out of eight on the two hundred yard circle. I think, sir, they probably died of Texas fever too."

The Second Law of Thermodynamics was saved again.

COMPLETION OF FIRST PHASE

The field tests on non-persistent gases were completed late in December, and Division 10 was preparing to leave. Malcolm Dole sent out a directive to the Division 10 team telling them how to organize and report the data they had taken.

RUSH TO GO TO PANAMA

There was a big hurry to transfer the NDRC group to Panama. Each would have to be inoculated against various tropical diseases that might be encountered. Late one morning I found Jim Pitts, an NDRC member from Northwestern University, slumped in a chair in front of the rented store we used for our offices, and he was in, or almost in, shock. His left arm hung limply down, and he had drops of blood in his left hand. He had reported that morning for his inoculations. The Bushnell staff doctors had the vaccine material, but the only needle they had was the huge type used to treat horses. There was so much hurry to get ready for going to Panama that the doctor used the big needle on Jim. Bob Brinton, in line behind Jim, fainted when he watched the doctor stick the needle into Jim and saw blood rushing out. From these reactions, they stopped using the big needle and postponed further inoculations until proper needles came in from Tampa, a day or two later.

Some NDRC members left by train. Dickinson's crew drove the 1942 Buick station wagon and the Caltech flat-bed truck. In their trip they slept one night in Mississippi, three nights in Texas, and one night in Arizona. By the next night they were in Pasadena.

VACATION

I received a week's vacation for Christmas, and I took the train to Atlanta. I had not been home for more than two years. My older brother Smith was an officer in the Transportation Corps aboard a supply ship in the southwest Pacific. My younger brother Richard was a Private in the Air Corps and was then based in England. My youngest brother Bill was ten years old. As Bill and I sat on the floor playing a game he had received for Christmas, my father remarked that the scene seemed like old times in the family. I gave my watercolor painting of the Gulf of Mexico to my mother as a Christmas present.

Switch to Persistent War Gases

MUSTARD GAS

In the fall of 1943, the agents tested by the Dugway Proving Mobile Field Unit were non-persistent, that is, they were gases in the air. They were poisonous to breathe, but eventually they were carried away by the wind, and with thorough mixing in air became harmless. During 1944 and 1945, the work was almost exclusively devoted to study of mustard, a persistent gas. Mustard gas was used during World War I, and although CWS had spent twenty years trying to find a more potent compound, mustard was still the most dangerous persistent gas then known. At room temperature, mustard is a heavy oily liquid, but it has some vapor associated with it. Mustard vapor, if breathed, destroys lung functions in a manner similar to and more severe than phosgene. The vapor penetrates skin, and a person wearing a gas mask can be killed by absorption of vapor through the skin. A trace of liquid of mustard on the skin is irreversibly absorbed in less than a minute, and it produces large crippling blisters. Heavy contact of the liquid on skin can be rapidly fatal. Ground or logs contaminated with mustard remained dangerous for weeks or more, as certain poachers in our area discovered.

NDRC DIVISION 9

By the first of January 1944, almost all members of NDRC Division 10 had left Bushnell, and they were replaced by a large group from NDRC Division 9, which was concerned with persistent war gases, mostly mustard gas. Another group of army officers also came to Bushnell. The NDRC Division 9 members that came to Bushnell were mostly organic chemists, and they were under the direction of Caltech Professor Carl Niemann. Arthur Pardee was an undergraduate chemistry major at Berkeley, a graduate student at Caltech, and was sent to the NDRC Division 10 group at Dugway. In Florida, he did some micro-meteorological work with me, did some work with Division 9 chemists, and became an expert in the construction of foxholes and protective bunkers used by the Japanese. In 1998, Pardee wrote two letters to me in which he included some of his recollection of this work. "We switched to vesicants (mustard) and wore protective suits as well as gas masks. The liquid 'gas' had a purple dye mixed in it, and sometimes we literally brushed our way through foliage dripping with the stuff when we collected samples."

Division 9 participants put out samplers for mustard before the tests, they picked up the samplers after the tests, and they chemically analyzed their product in the laboratory in Bushnell. Also, they measured mustard gas concentration by a long paper tape treated with chemicals to give a color change when mustard gas was present. A clock mechanism unrolled and re-rolled the tape such that only a small length was exposed to air at one time. This instrument was called a Brown tape recorder.

HEADQUARTERS

For the new work, the Division 9 group would need all of the Division 10 office space to use as their chemical analysis laboratory. Captain Nolen offered me a desk in the large store building he had rented for his headquarters. Middle-aged Dr. Duncan MacRae, a civil service employee with the Chemical Warfare Service, had a desk in the room. He worked for Nolen, and apparently was especially useful to him. MacRae loved the out of doors, fishing, and hunting. There was excellent fishing in the vicinity of Bushnell, and MacRae made good use of it.

Second Lieutenant William Ironsides and Second Lieutenant Alan Englander were Air Force meteorologists, and they were stationed indefinitely in Bushnell. They were not there to forecast the weather. Under the new arrangement these two Lieutenants worked under my direction. Ironsides stayed in the office and helped to organize and interpret data, especially the micro-meteorology data my group produced. Alan worked directly with me in the field and worked with the data in the office. I never met the meteorologists who forecast the weather and were stationed in Dade City.

Later, Robert Merrill from the University of Chicago joined my group. He was an expert in mathematics and statistics. He played an important role in our interpretation of data, and sometimes he joined us in the field work.

Captain Nolen assigned a sergeant to me, but I have forgotten his name. He had been in the chemical warfare service for several years before the war started. He learned to operate all of our micro-meteorological instruments, and he supervised the three or four soldiers temporarily assigned to our group. During 1944 and 1945, several soldiers would be assigned to Bushnell for a period of about three months, and then they would be sent on, overseas, presumably. The micro-meteorology work required some hard labor, in particular moving and recharging the heavy-truck storage batteries that we used to drive the vacuum cleaner motors to aspirate our temperature-measuring devices. Other heavy equipment had to be moved in and out. My sergeant was invaluable in directing the work of these soldiers, and he helped keep the instruments in repair. I was talking with the officer who assigned the new soldiers to various units, and I jokingly suggested he send us a scientifically trained draftee. Somewhat later he proudly told me that he had assigned a meteorologist to my operation. When the Private arrived, I started discussing our project with him, but he seemed to understand none of it. I asked him what work he had done before getting drafted, and he said he was the meter reader for the gas company.

LIVING QUARTERS

Upon arriving at Bushnell, I rented a room in the sheriff's house. Later I moved from my rented room in the sheriff's house to a rented room in the mayor's house. The mayor was the pharmacist of the town. His wife Naomi was plump, cheerful, friendly, and concerned about my social life. I was invited to a party at the mayor's house one night. The guests were mostly middle-aged, like the hosts, but there were some young people there. One of the girls at the party and I became friends, and we had several dates over the next year (I have forgotten her name). The mayor's party mostly consisted of sitting in the dark outside talking and drinking orange juice, that is, orange juice fortified with pure ethyl alcohol illegally provided by the pharmacist. The mayor's wife was particularly invigorated by the orange juice; she sang and danced and, unknown to her, snagged a large hole in the seat of her lightweight summer pants. The mayor growled, "Set down Naomi, your bottom is out." The party continued, but Naomi was not put down.

Enlisted men lived in a camp of tents just outside the town, and an army field kitchen served meals to them. Chemical Warfare Service officers rented housing in town, as did the civilian NDRC workers. Second Lieutenant Alan Englander and I rented an old house in Bushnell, which had a hall down the middle, four rooms, and a large enclosed back porch. It did not have a shower, but it had an old fashioned bathtub mounted on four legs. The house had a large number of cockroaches.

We cooked some meals, but often we ate at the local restaurant. Bushnell is close to the center of Florida, north and south, east and west. Its culture and its cooking were deep south. I had eaten for two years in California, and I came to dislike much southern cooking. In the south, fresh vegetables, in the presence of the salted fat hog-belly, were typically boiled for hours and sometimes simmered all night. We ate at the local restaurant in Bushnell. (I forget whether there was one restaurant or two in Bushnell). Fried chicken, country-fried steak, and fried veal cutlets were typical meats. The vegetables were all overcooked in the presence of solid pork fat. I sometimes took my plate of served dinner, shook a heavy dose of black pepper at spots in the food, under lifted portions of the mashed potatoes, and in turnip greens and meat, and then spun the plate. In this case I never knew as I took a bite whether it would taste like black pepper or hog fat. That provided variety.

MEDICAL FACILITIES

Captain Bert Kribben and Captain G. H. Mangun were medical doctors and members of the Medical Division, CWS, and they were assigned full time to the Bushnell operation. At the edge of town they rented a large old two-story Victorian house for their offices and as a hospital. Their staff included male army nurses and other assistants. In January 1944 I went to them for my first physical examination, according to my agreement with Captain Nolen. One of the doctors told me that I could expect to live between ages thirty and forty. I told him that was what several other doctors had said.

MICRO-METEOROLOGICAL TREE TOWER

I had a long talk with Jake Nolen. I pointed out that in the fall we had taken micro-meteorological measurements in the meadow and in the forest, but the measurements in the forest were not meaningful since the observations only went up to five-meter (16 ft) height while the forest canopy was about twenty-meters high. I asked if we could obtain a steel meteorological tower that would go up above the treetops. He said that our operation had been assigned such a tower, but it had been erected and used in Dade City by our weather forecasters there. There was no chance of our getting another tower. I came back the next day with a new proposition. I had noticed some tall pine trees in the hammock, and some of these went above the canopy of the hardwood forest. Could we strip one of them and put in steel climbing rods like those on telephone poles? He promised to check up on our prospects there. A day or two later, Nolen said he had talked to the forester of the Withlacoochee Land Use Project, who had flown over that forest many times. He had confirmed that some pine trees grew tall in the forest, and he had in mind one that was about half a mile (0.8 km) away from the test area. Captain Nolen set up an appointment for me to go out with the forester to look at that tree.

I rode out with the forester in his pickup truck. The map on Figure 4.1 shows a road crossing highway 471 just below the label "471" on the map. This road was the main entrance to the test areas, it crossed to the north of the forest test areas, and it led to the north side of the meadow. The forester drove down that road until we were well inside the hammock. We stopped and worked our way through heavy underbrush for about a hundred yards until we came to the tree, which stood straight up and went above the main forest trees. The first live branches of the tree began about two-thirds way to its top. We agreed that those branches should be cut, leaving stubs one or two meters long to hang the instruments on, and steel climbing rods would be placed up the tree from the ground. I was not there when the work was done, but I came out as soon as I heard the job was finished. They had cleared a narrow trail through the underbrush from the road to the tree; each day as we went down that path we brushed aside the spider webs that crossed the path. I could easily climb to the top of the tree by use of the climbing rods and the stubbed-off branches. Figure 4.2 gives a view of the forest above the canopy and a view the forest at ground level.

Lieutenant Alan Englander, my sergeant, an electrician, four soldiers, and I brought our equipment to the tree, and we proceeded to mount it. We had a large vacuum cleaner powered by two heavy-truck storage batteries. It aspirated six temperature-measuring devices through vacuum-cleaner tubing, and each station had a vacuum-cleaner tube and wires to the ground. These temperature-measuring devices were 0.3, 5, 10, 15, 20, and 25 meters above the ground and placed at least one meter away from the tree trunk. The new Esterline-Angus meters had three pens and could record three separate measurements, and two of these meters simultaneously recorded all six temperatures. As an essential step in our work, we used a high precision Leeds and Northrup potentiometer at ground level to calibrate the temperature-measuring devices twice a day, but these measurements required two operators, one in the tree and one on the ground. Another pair of meters recorded the wind direction at twenty-five meters above the ground and wind speeds at 23, 15, 10, 5, and 2 meters above the ground. We had a small rain-proof wooden shelter to protect the recording meters and potentiometer. Our "meteorological tower" is shown in Figure 4.3. Note the air tubes and wires fastened to the trunk of the tree, the climbing rods, the soldier partway up, and instruments attached to rods or to stubbed branches of the tree.[#]

During field tests, we made simultaneous measurements in the forest and in the meadow. Typically, Robert Merrill and I, with the help of a couple of soldiers, operated the forest station before and during each test, and Alan Englander, the sergeant, and another pair of soldiers operated the meadow station. On some occasions I went to the meadow station, and the four of us, including Alan, the sergeant, and Bob Merrill, switched jobs from time to time.

AIR STABILITY IN THE FOREST

With clear skies, the meadows near our test site had strong temperature inversions at night, and during the day they had unstable "lapse rates," that is, the temperature decreased with height above the ground. This pattern was typical and widely recognized with respect to grass lawns, cow pastures, barren hillsides, and desert soils. After we converted the tall tree into our micro-meteorological tower, I wanted to make temperature profile measurements and wind speed measurements up through the trees twenty-four hours a day, whether tests were being conducted or not. I was curious about the magnitude and timing of temperature inversions in the forest and of the relation between meadow and forest. I made arrangements with Captain Nolen to have recharged batteries delivered twice a day at both the

[#]Dugway Proving Ground Special Report No. 35, Project E 7a-2. "Micro-meteorology of the woods and open areas within the Withlacoochee land use project," 30 pages plus 20 pages of figures, by Harold Johnston, Div. 10, NDRC, assisted by Arthur Pardee, Div. 10, NDRC, Second Lieutenant Alan Englander, AAF, Second Lieutenant William Ironside, AAF, September 1944.

Figure 4.3. Micro-Meteorological Tower, Withlacoochee Hammock, Florida, 1944–45. Picture by Army photographer, July 1944.

meadow and forest stations. The stations recorded temperatures unattended during the night.

We recorded continuous measurements throughout the months of February and March of 1944, a period in which there were few leaves on the deciduous trees. Also, we made continuous measurements during the months of May

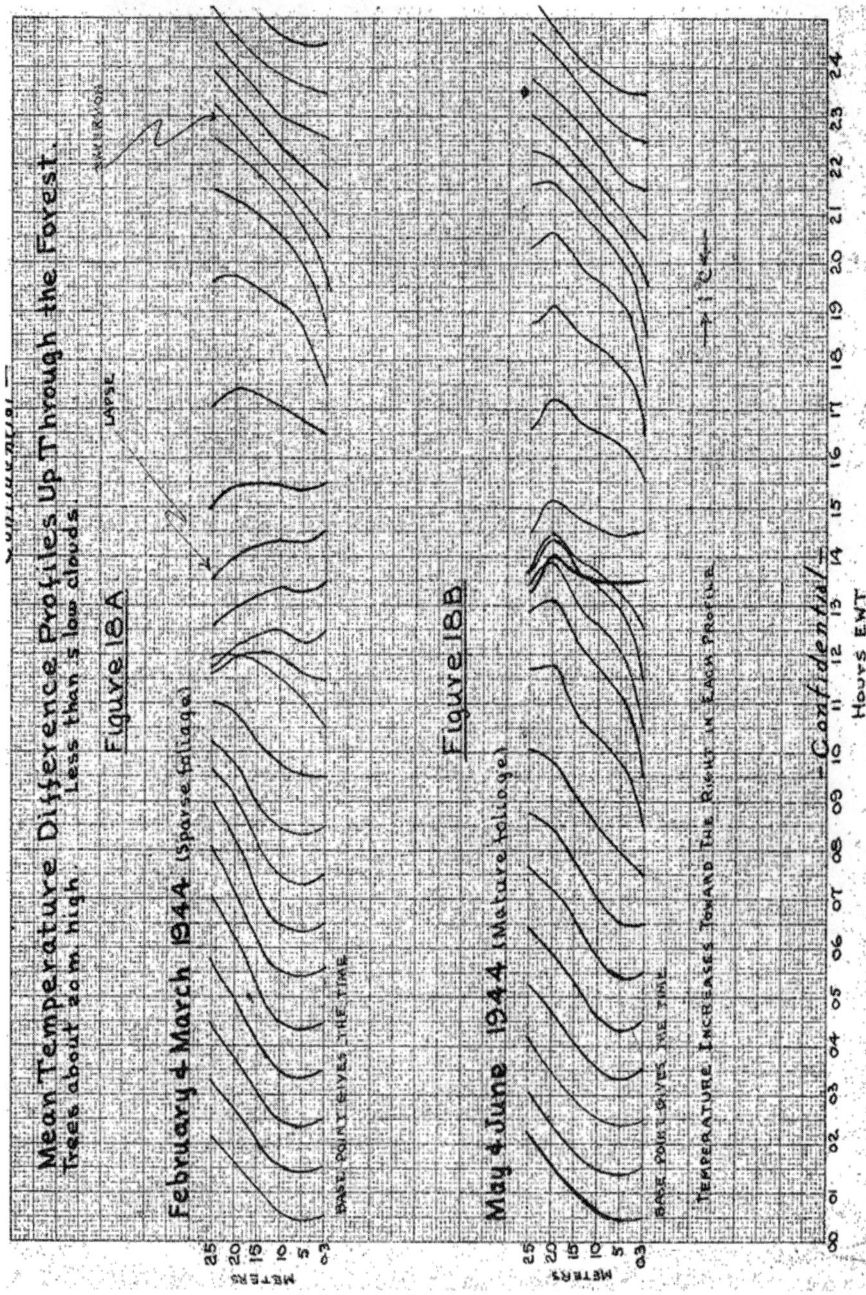

Discovery that air in a dense forest has a strong temperature inversion throughout almost all of the day and a weaker temperature inversion during the night. The temperature profile above the canopy was quite similar to that above the meadow.

Figure 4.4. Figure prepared by army personnel, 1944.

TWENTY-FOUR HOUR TEMPERATURE DIFFERENCES IN WITHLACOOCHEE MEADOW AND DENSE FOREST, May–June 1944.

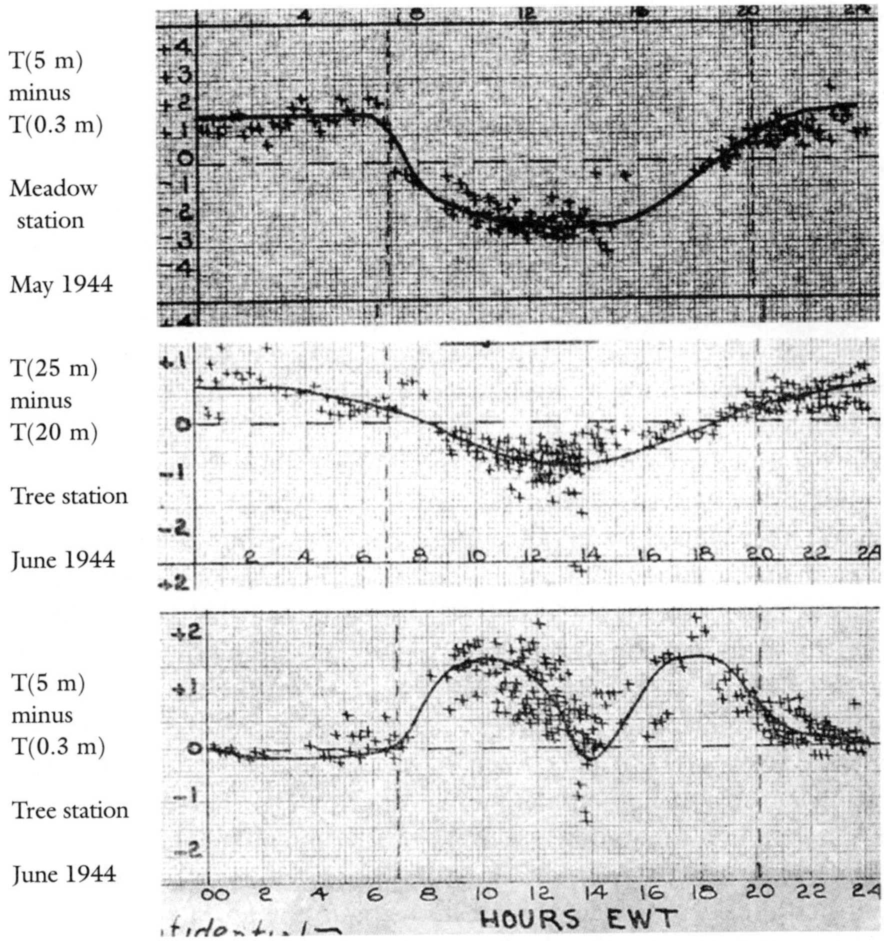

Discovery that air in a dense forest has a strong temperature inversion throughout almost all of the day and a weaker temperature inversion during the night. The temperature profile above the canopy was quite similar to that above the meadow.

Figure 4.5. Figures prepared by army personnel, 1944.

and June 1944, which was at the time of maximum density of foliage in the forest. From our averaged data, we prepared vertical profiles of temperature, but we rejected data when the sky was more than half-covered by clouds. Figure 4.4 shows vertical profiles of our temperature measurements taken twenty-four hours per day, 0.3 to 25 meters above the ground, for February and March, and for May and June. We drew significant conclusions from these measurements. The temperature differences between air in the forest canopy and above the canopy showed the same pattern as that observed in the meadow over the course of twenty-four hours: inversion during the night and unstable lapse rate through the day Figure 4.5. From 0.3 to 15 meters above the ground surface, the pattern was totally different between the meadow and the forest. During May and June the forest air showed a temperature inversion, which reduces vertical mixing, every hour of day and night. During February and March, air in the forest had an unstable lapse rate from about 1130 to 1530 Eastern War Time, and it had temperature inversion at all other times. In the lowest five meters above the ground during May and June, a strong temperature inversion built up by an hour after sunrise, but this daytime inversion decreased to zero at about solar noon. Another strong inversion built up during the afternoon and evening. The daytime inversion in the forest was a mystery at first.

We explained this unusual pattern in the following manner: sunlight absorbed by the leaves warmed canopy air to temperatures higher than the air above to produce an unstable lapse rate above the canopy during the day, as in the meadow. This sun-heated air in the canopy was warmer than the shaded air in the lower forest, which constituted the daytime inversion in the forest. During the early morning and late afternoon, because of the slant angle of the sun, foliage cut out almost all of the sunlight from lower levels of the forest, and the inversion was large. When the sun was nearly overhead, some sunlight penetrated to the ground to reduce or break up the temperature inversion during the middle of the day. The data for February and March, when the canopy was at its yearly minimum, showed much less day-time temperature inversion than that observed in May and June. We formulated the rule that the daytime inversion in the forest would be greater the denser the canopy, that is, it would probably be highly important in tropical jungles. With rare exceptions, air in the dense forest was stable against vertical mixing, all night and all day (Figure 4.4). Also, wind speeds in the forest were lower than wind speeds in the meadow. Poison gas clouds would persist in a dense forest much longer than in open areas.

The daytime temperature profiles were easy to interpret, as done above, but the nighttime profiles did not have a simple interpretation. Sunlight impinges on the forest as a directed beam, some of it penetrates the canopy, especially with almost overhead sun. Day and night, heat radiation is emitted by all surfaces and in all directions, and radiatively cooled leaves chilled adjacent air, which sank toward the ground. The net effect in our examples was that there was a minor temperature inversion throughout the forest during the night in all cases and a strong temperature inversion throughout daytime when there was a dense canopy of high level vegetation.

We ran into one difficulty during our overnight recordings during May. The Esterline-Angus meters recorded in red ink. Well into our program, we found one or more of the ink traces went dry in the early evening. We were doubly careful to see that we properly filled the ink reservoirs, but we still lost some of our records. I drove out one night, brought along a strong flashlight, and set up a chair about two meters from the recorders. For a few hours I heard owls and other night noises, and then I heard a rasping sound in our box of recording meters. I turned on the flashlight and found in the recorder the biggest cockroach I had ever seen. It was as big as a cigar and about that color, twelve to fifteen centimeters (5 or 6 inches) long, and when I opened the case and tried to catch it, it flew away on big whirring wings. It drank the red ink of the recorder. Later, someone told me it was a "blabberous cockroach." The next day my sergeant took a soldier skilled in carpentry to the site, and I went along. The carpenter installed two tight wood-and-screen doors, and he stopped other possible holes with putty or by covering with a tacked-on piece of screen or wood. He did a good job, because we had no more trouble with ink-guzzling cockroaches.

MAC

One day Second Lieutenant Robert McWhorter (except for the "Mc," this was not his actual name) showed up in Captain Nolen's office with papers stating the he had been assigned to the Bushnell CWS unit until further notice. After Captain Nolen interviewed the Second Lieutenant, he could think of no suitable job him, unless, perhaps, in the micro-meteorology department. Nolen called me in to meet the newly arrived officer. McWhorter was a big guy who had red hair, blue eyes, narrow steel-rimmed glasses, and freckled face and hands. He appeared to be two or three years older than I. Captain Nolen inquired if McWhorter could work in our group. I agreed to have him join us. When it appeared to McWhorter that he would be directed by a civilian, he strongly protested, but Nolen explained that I was in charge only between the alert signals a half-hour before and at the end of the test. At other times and other places I had no authority over him. Obviously McWhorter did not like this situation, but he let me explain our meteorological program, and I went over what we did for each test. I mentioned that we had two enlisted men who did the heavy work such as moving and recharging our batteries.

During the next test, Alan Englander and I took Mac, that is, McWhorter, to the site and showed him how to read the instruments, but he had difficulty understanding how to do it. During later tests, he came out in his own jeep. Mac ordered the two enlisted men to do petty things for him. This went on for several tests, and the new Lieutenant became demanding and nasty to the two soldiers.

One day, there were only Mac and I at the forest station, and it required two people to calibrate the instruments before each test. I asked Lieutenant McWhorter to climb the tree and throw the switches as I instructed him, and I would make the measurements with our potentiometer. He bluntly refused to climb the "dirty" tree. He was wearing a clean officer's uniform, and there was some residual pine sap where the branches had been cut. Other officers in the field dressed in work clothes, and Lieutenant Englander readily climbed the tree. Time was pressing, and I said I would climb the tree if he would read the potentiometer, which I had showed him how to operate. I climbed up to the first station and asked him for the first reading on the meter. He fumbled and fussed and finally shouted that he could not make the machine work. We argued. As he left in his jeep, he called me, among other things, a sissy draft dodger. I hid my ailment whenever possible, and my strength had improved so much that I could usually get away with it. Throughout Mac's stay at Bushnell, he resented the fact that he was being directed by a civilian, especially one younger than he and one conspicuously non-athletic. Aside from Mac, I got along well with the military personnel.

The next day Captain Nolen received an official telegram stating that Second Lieutenant McWhorter had been transferred to another assignment. (I suspected that Mac already knew about the transfer on the day we had our argument).

KISSING BLISTERS

In the summer of 1944, I reported to the army hospital for my appointment to have a quarterly medical examination. I was told that an emergency had arisen, and a doctor would not be able to see me until much later. I waited on the shaded front porch of the old house that was then the hospital. Later, a large open army truck stopped in front of the hospital, and it was followed by two jeeps. About eight or ten nude soldiers stood on the back of the truck, holding the side rails. Some were draped with a towel. Soldiers from the jeeps and nurses from the hospital moved to the truck, helped the men climb off, and escorted them one by one into the hospital. The first one in was a large man of athletic build; he had a series of blisters from under his left armpit down to his left knee. Most of the blisters were the size of a half grapefruit, some were bigger and some were elongated, and from the weight of the fluid they sagged like breasts of old women. The next was a short, plump young man with a huge blister hanging down from each buttock, which rubbed against each other as he walked. From this effect, a nurse called them "kissing blisters." The parade continued of soldiers with large sagging blisters over various parts of their bodies. I left a note postponing my medical appointment.

The army had exploded on the ground an array of bombs filled with mustard gas, and then a group of soldiers with gas masks and full protective garments marched into the contaminated area and stayed there overnight. Later they marched out, decontaminated themselves, and removed their garments. The garments had failed to protect some soldiers against crippling blisters from the mustard.

Another time at the height of summer, I was waiting on the hospital porch watching a soldier digging a trench while wearing a full protective outfit, including gas mask. Someone told him the time was up. He stopped digging, took off his rubber boots, and poured sweat out of them. He then came inside to change the rest of his clothes and presumably be examined by a doctor.

One of the doctors asked me to participate in a test he was carrying out. He had a bottle with a small amount of mustard gas in it, dissolved in some solvent. He put one drop on my left arm and told me to record what happened over the next several days. A hemispherical blister formed over an area about the size of a silver dollar. It was painless, and the stretched skin was smooth and tight. After a few days the blister dried and cracked releasing fluid, the skin came off, and there was a sore for many days. The sore healed, but even now, year 2001, I have a constellation of small red spots on my arm where the mustard blister had been.

PRECISION BOMBING

Our project had some difficulties with the delivery of mustard gas bombs by airplanes. On one occasion the plan was to drop six mustard bombs in quick succession. The bomber moved on its course, approached the delivery point, and its bomb bays were opened; but the crew detected something wrong, they did not release the bombs, and the pilot took the plane on a wide circle for another pass at the target. As they approached the target, the pilot called out to the bombardier to prepare for the release of the bombs. Then the bombardier found that the bombs were not there; they had fallen out somewhere on the large flight circle. No one ever reported the landing of these bombs. Presumably, even now, six rusting bombs full of mustard gas lie buried in some swamp in the Withlacoochee area.

Captain Nolen explored the possibility of getting better bombers for his project (anyone in the room with its open offices could hear his forceful clear voice when he talked on the telephone). The Air Force proudly replaced the old aircraft and their crew with the latest bombers, their new precision bomb sights, and skilled pilots. Alan Englander reported that, in the officers mess, the new pilots were boastful and arrogant, in general, and especially about their precision bomb sights. Captain Nolen scheduled an airdrop experiment. On that day I happened to be working at our meadow station, and someone else took care of the forest station. Soldiers had whitewashed a tree to designate the target. Our micro-meteorological station was placed to the east of the road through the meadow. The field officer had the trucks form a double line along the road, and although the trucks were a half-mile from the target, the officer required all personnel, including about a dozen NDRC men, to crouch behind the trucks, away from the target. The officer appeared to be excessively cautious. It was a hot sunny day, and many of the soldiers removed their shirts to get some sun. The bomber was flying west to east over the target. We heard the plane approach from afar and then nearer. It passed over our group of trucks. We heard a short screeching sound, and the five hundred pound (227 kg) bomb of mustard exploded in the top of a pine tree just to the east of us. The soldiers scrambled to put on their shirts and gas masks. We noted a slight smell of mustard, but the wind carried most of the mustard away from us. As one NDRC person stood up, he said, "I wouldn't call that precision bombing. They missed us by a hundred yards."

* * *

This chapter covers the period through September 1944. The NDRC Division 10 expedition to Panama spanned the period February through September 1944, and the tests in Panama are the subject of Chapter 5. Chapter 6 continues our experiences in Florida from the end of Chapter 4 to the end of the war.

CHAPTER 5

JUNGLES

I did not go to San Jose Island, Panama, and this chapter is built up mostly from quotations written in the 1940s by those who did go there and from pictures taken in 1944. I start with two quotations from W.A. Noyes, Jr., Head officer of Division 10 of the National Defense Research Committee (NDRC).

> Although the Florida Withlacoochee area provided isolated semi-tropical forest and meadow for the chemical warfare tests, it was not equivalent to the jungles of the southwestern Pacific, and it was not far enough removed from civilian habitation to be suitable for large scale testing of weapons. Colonel R.D. McLeod, representing the Chemical Warfare Service (CWS), and Dr. Carey Croneis (University of Chicago), representing the National Defense Research Committee (NDRC), explored the Panama Canal Zone and identified the uninhabited San Jose Island about sixty miles (97 km) out in the Pacific Ocean from Balboa as a site that had all the desired features. San Jose Island was selected, Brigadier General E.F. Bullene was commanding officer, and Dr. Francis Blacet was designated as head of the NDRC Division 10 group. [W.A. Noyes, Jr., page 321]

Odyssey

> The entire expedition sailed from New Orleans in February 1944, and finally arrived in the Canal Zone after a lengthy trip which covered most of the Caribbean Sea not once but several times. [W.A. Noyes, Jr., page 321]

Twenty-one civilians in NDRC Division 10 boarded a troop ship in New Orleans in February 1944, which joined other ships to form a convoy that sailed at about six miles (10 km) per hour toward the Panama canal, a trip that should have taken between one and two weeks to complete. However, the ship crossed the Caribbean Sea several times, as Noyes said, and the confusion and amazement of the NDRC men were expressed in letters written home by John Otvos and Bob Mills.

JOHN OTVOS ACCOUNT OF TRIP

In early February 1944, members of Division NDRC assembled in Jackson Barracks of New Orleans, which was primarily occupied by army soldiers. John Otvos wrote letters to his parents, which I quote in part here and more later. I omit salutations and farewells:

> **Feb. 12, 1944.** I have been so busy the last few days that all I have had time to do was think of you. Writing was out of the question. I can't tell you much about anything anyhow, because it would be censored. I am having a good time, though, so far, and enjoying myself very much. The food is good and the company just as enjoyable as I expected ... Here it is Sunday already, and it seems like only yesterday that I left home. The last couple of days have been very cold, just as it was in Florida in December ... Yesterday we stocked up on a lot of army clothes, including shirts, pants, socks, work suits, and even shoes. [But the army garments included no emblems.]

For this book John Otvos submitted an account of some of their activities on board the ship:

> It was during this period (early February 1944) that our group got organized, without any planning, into two sections: chess and bridge. For most of the next month we had no duties, and we played these games constantly, sometimes on our cots, sometimes on the stairs, and in any other sheltered spots we could find. Bridge scores for individuals were cumulative although partners and venues changed. The chess group had tournaments and a chess ladder with challenges. As I recall the groups never intermingled. I was a chess person and so were Phil Hayward and Ted Gilman. We had some chess books with us, which I'm sure we bought in New Orleans.
>
> There were army personnel who shared our quarters in Jackson Barracks in New Orleans who became quite curious about our group and kept prying us with questions. Phil Hayward got rid of them by saying we were studying "astrobotany."

Bob Mills and Rene Scott (later Rene Mills) exchanged correspondence over the entire trip to and from Panama. In one letter Bob said they were "languishing in the barracks on the gulf coast," which Rene interpreted to be New Orleans. In a letter, Rene said to Bob:

> Got a letter today-hooray! It's wonderful to hear from you. Soon after I'd read it I saw R.G. [Roscoe G. Dickinson]. He was wandering around muttering and looking unhappy in general. He asked me if I'd heard from you, and then with a gleam in his eye he asked if I knew where

> you were. For a minute I was rather worried, but then I discovered he was just asking for his own information. He asked about the date the last letter had been written and wondered if there had been any hints of your whereabouts. He is confused about your location and wanted me to help him out. Thought you would be interested that the top guys are not very well kept up to date.

They were delayed for ten days in New Orleans waiting to board the ship. John Otvos and Ted Gilman bought books and equipment for navigation by the stars. The group enjoyed a dinner tour of the city under guidance of Bob Brinton and Dave Volman. At Chalmette harbor on 19 February 1944, they boarded the 10,000 ton troop ship named William Everts, which was equipped with one five-inch (13 cm) cannon, but they were immediately stopped in New Orleans. The army unloaded all the cargo looking for munitions in the hold, which was not permitted on a troop ship. The army found no munitions and reloaded the ship, but this inspection delayed departure by five days. They left New Orleans and sailed out into the Gulf of Mexico on 24 February 1944.

> **Feb. 28.** This writing from a ship is a bit strange. I won't be able to mail this until we reach our destination which may be a couple of weeks off. But rather than wait all that time, I think I'll keep a diary and add to it when the spirit moves me. I have been on the ship for 10 days already, but we didn't leave port until 5 days ago because of some cargo difficulties. However, during that time we were not allowed to leave the ship so we were as good as at sea. There was just one meal after another with a little sleep in between. Really if it weren't for the fact that we are anxious to get to work and do what we are supposed to do, it would be a very enjoyable vacation ... Not being enlisted men, we have no kitchen and other duties, and not being officers we don't have to be responsible that these duties are carried out. So we read and play chess and bridge all day. The weather has been perfect since we have been sailing ... The ship just doesn't sway at all.
>
> Twenty-one of us live in one ward room about the size of our living room, if not smaller. The bunks are three deep and I have one of the top ones. There's more air up there but the whole is really much hotter than it should be, and as soon as the blackout is lifted in the morning everybody rushes up to the deck and stays there all day ... We are supposed to follow all the rules prescribed for officers...
>
> Naturally, we are sailing in convoy and when I look around and see the other ships and the occasional airplanes, it's impossible to worry about submarines. Now and then at night we have a movie under rather

> primitive circumstances ... And every now and then we're entertained by some Latin American music played by some of the soldiers. It all makes for a pleasant cruise and in peace time it would cost a lot of money.

The ship landed in Guantanamo, Cuba on the second of March, all waited aboard the ship a couple of days while another convoy was formed, which departed on March fourth.

> **March 5.** Ever since my last exam at the end of May [his final examination for his Ph.D. degree] I have been doing mostly routine stuff ... and manual labor in Florida. And now for a month I've been sitting. This trip is taking a lot longer than thought and its a shame when you realize it could have been done in 12 hours. One of our boys paraphrased Churchill, 'Never before have so many done so little for so long.'
>
> We made a crude but adequate sextant out of two mirrors and I bought a book about navigation. Using these and a World Almanac we can tell from the sun or the stars exactly where we are on the sea. It's fascinating business and makes the trip more interesting.

John told me that the ship posted the location of enemy submarines on a map, but it did not reveal the location of the convoy.

> Ted Gilman and I obtained material to make a small sextant with which we could measure the altitude of the sun. The maximum altitude together with the sun's declination gave us our latitude, and the time the maximum was reached together with the equation of time gave us the longitude. The CWS Captain obtained the exact time for us from the bridge once a day. In that way we could plot our position daily and see how close the submarine sightings were from us in the Gulf of Mexico.

On a couple of occasions the submarines were fairly close.

The convoy proceeded for three days from Cuba to San Juan, Puerto Rico, where they arrived on March 7. The NDRC group had several days of shore leave. John Otvos estimates that one hundred or more soldiers were loaded onto the ship, and two hundred goats were brought on and tied to the railing around the deck, thereby usurping the best locations on the ship. They sailed from San Juan on March 12, proceeded, not to Panama, but back to Guantanamo, Cuba, where they stayed for two days, and then on March 15, they headed for Panama.

BOB MILLS' ACCOUNT OF THE TRIP

A typical envelope had "Robt. Mills, APO # 827, c/o Postmaster, New Orleans, La." in the upper left-hand corner, and "Rene Scott, Crellin Chem. Bldg., Caltech, Pasadena, Calif., U.S.A." as addressee. A round stamped message read: "Passed by Base 0009 Army Examiner," and the envelope had been slit along the left side and resealed with tape with the words "Opened by Army Examiner." There are forty-three such letters written between February 13, 1944 and September 20, 1944. In this period, Rene Scott wrote forty-five letters to Mr. Bob Mills. Bob's letters revealed nothing about the military tests being carried out, but he gave a fairly clear picture of traveling conditions, living conditions, occasional recreation, and things about the people around him. Without exception, each of Bob's letters contained a section about baseball. Rene's letters gave news and gossip about what graduate students, secretaries, and young faculty members were doing at Caltech. Without exception, each of Rene's letters contained a section about baseball. Each prodded the other to send more letters. These letters have none of the fury and seriousness of Sam's letters to Helena (Chapter 3). Bob and Rene had been more than just friends for several months, and during the month of January 1944 between Bob's tour in Florida and his departure for Panama, they came to a deep understanding and commitment. (As they note in their letters, some of this occurred while they were in my Buick convertible). Bob joked, drew droll tiny cartoons, and humorously misused the English language. The beginning and conclusion of each letter expressed affection and longing in widely varying ways from one letter to another. Here, and again later, I quote small portions of some of Bob's letters, in general omitting expressions of affection and discussions of baseball:

> **Feb. 13.** Things is grim! By now I am well situated in a barracks somewhere on the Gulf Coast—enjoying army food, clothing, bed, and company. Just call me G.I. Joe ... I am writing this from the officers' club where things are 'just like home.' Thank god you can't see me in this uniform.
>
> **Feb. 22.** We had mail call today & it was a real treat when I heard my name called & was handed a pale blue envelope. Now I know what they mean when they say write to the boys in the service ... The army has done a fairly good job of butchering things up, but I can almost say for sure that this will be my last letter for a month ... I will write you very often—even though I can't mail the letters until later. And if you do the same, I shall love you for it when I spend my first day there (near

there) reading all of your accumulated letters. If it doesn't make sense please don't read the above. Anyway you can see what state I am in. Ooops, I'm not supposed to tell you what state I am in. Censored …

Remind me to donate to the Red Cross at the next possible opportunity. It has treated us swell—with gifts of all kinds.

Thought you might like to know that by now most of the guys have blossomed forth with fine sunburns. It is especially noticeable on the gang of palefaces from Northwestern.

March 5. No doubt you must realize that we are still riding the waves & have today started our third week before the mast. Bill Gwinn is still seasick … Fortunately, I haven't missed any meals yet—one way or another—and since my last letter the ship has picked up quite a pitch and roll—which ain't exactly pleasant—but such a rough sea isn't pleasant for submarines either.

We touch land tomorrow [San Juan, Puerto, Rico] for I don't know how long—and I am hoping madly that I will be able to mail you this stack of letters.

Honest, R. Scott, all I really want to say in this letter was that "I went to Church today." They have services every Sunday on the ship & although the attendance is small, the singing is always loud (and sad I might add).

I played another game of chess today—it lasted for four hours & ended in a stalemate.

When we are docked or in port we get to see pictures on deck.

March 11. I miss your news items, bright sayings, quips, and literary dim views. Do you, Betsy, & Judy still beat a path to the Stuft Shirt for a Gibson or so?

Martini + onion = Gibson + olive.

Well, Rene girl, my nose has just peeled for the fourth time, and my face resembles a traffic light that registers STOP all the time.

At the last stop I bought a book on art and drawing—so that I may improve my sketches. (the fact that it is just chock full of nude female models, etc. had nothing to do with my purchase).

Will you please write me & let me know if the war should happen to end? Or anything else like that which I should know.

March 14. Due to an unexpected landing & still more unexpected shore leave, I was able to post some letters to you.

A lot of exciting things happen on the transport but the gov. is so touchy about letting me tell you. For instance, you'll probably find a big hole where I mention the gun practice we had yesterday. They fired anticraft & 5 inchers & I might add my ears are still ringing.

More misery—meals have been cut to 2 per day & this causes untold suffering to one R. Mills.

March 18. I am still being tossed about by this same ship, and at present I am cooped up in Ward room #4 stewing in my own juices, for it is hotter than the Rancho on the fourth of July with an R.H. [relative humidity] of 3 figures.

Tonight will be my last on this barge and tomorrow we dock near the destination ... On the day after tomorrow, Monday, it will be six weeks since Bill and I left. I wouldn't have missed this cruise for anything, and although time would hang heavy at times, there were always sunbathing, pictures, chess, poker, books, drawing, bridge or the old standby, bull sessions. I believe we must have discussed (& cussed) about every topic under the sun, and usually the conversation drifted around to girls at which point sex would rear its ugly head.

The food has been par excellent during the entire trip & leads me to the realization that it must be tough being a civilian these days.

March 22. Wed. morning. On Sunday night we finally disembarked & after a quick supper (on shore) we were driven to a civilian barracks ... the place is pretty nice & it looks as though we'll be stuck here at least another week. There is nothing to do but sight-see.

Met Captains Stone and Quate ... I ran into a fellow on the boat who is from back home. He is C.O. of a gun battery here & has invited me to spend tomorrow.

When I got here I found 24 letters ... I saved your 5 epistles 'till last.

John [Otvos], Bill [Shand], and several others are planning a big sight-seeing trip tomorrow, but I guess I'll be content with just Christabel, Panama City, Balboa, Tivoli, & Corundu.

They arrived at Colon on the Atlantic Ocean side of the Panama canal, went through the Panama canal, and reached Balboa on the Pacific Ocean side on March 19. San Jose Island did not have a harbor deep enough to accommodate the troop ship, which anchored off shore. Officers, soldiers, NDRC personnel, goats, equipment, machinery, and cargo were loaded onto a landing barge and brought ashore, Figure 5.1.A. It took many trips of the landing barge to unload the ship. The NDRC group reached their destination on 28 March 1944, thirty-eight days after leaving Chalmette, Louisiana and almost exactly one hundred days after Jim Pitts went into shock from being vaccinated for yellow fever by a horse needle in Bushnell, Florida, because there was such a rush to get off to Panama. Their travels were:

Figure 5.1.A. Temporarily beached landing barge, which was used to unload lumber, machinery, weapons, vehicles, goats, etc. San Jose Island, Panama, 1944.

Figure 5.1.B. San Blas Indians' (the servants) quarters, located just beyond hill cleared to build the army camp.

Photographs by Bill Shand. Provided by Mrs. Rene Mills.

Leave Chalmette	1515	19 Feb 1944
Leave New Orleans	0800	24 Feb
Arrive Guantanamo	1000	2 Mar
Leave Guantanamo	0600	4 Mar
Arrive San Juan	0800	7 Mar
Leave San Juan	2030	12 Mar
Arrive Guantanamo	1545	14 Mar
Leave Guantanamo	1300	15 Mar
Arrive Colon	1000	19 Mar
Arrive Balboa	2000	19 Mar
Arrive San Jose Island	1100	28 Mar

I give quotations to illustrate both sides of heated disagreements about the development of the San Jose project. One side was Washington, DC, and the other General E.F. Bullene, commanding officer on San Jose Island.

WASHINGTON DC VIEWPOINT (W.A. Noyes, Jr., page 322)

> Upon arrival at Balboa it was evident that the camp site and the various facilities on San Jose Island were not ready, and it was April before the group moved to the island and the first experiments were carried out.
>
> The timing was unfortunate because the advent of the rainy season proved a severe handicap from early in May until the time the Division 10 group left. There were many trials and tribulations, arising partly because of the conflict between emphasis on housekeeping and military matters and emphasis on the technical aspects of the expedition. At best, conditions were bound to be difficult because of the need of making roads, providing adequate housing, and obtaining all the necessary supplies. Nevertheless, much useful information was obtained. In August and September 1944 several attacks were performed under operational conditions with the aid of the 6th Air Force on a scale which, as far as we are aware, was never equaled in gas experiments during this war. The data obtained were more complete than those ever obtained before. A large amount of credit for organizing the nonpersistent phase of the program is due to Dr. Blacet, but the entire group with him performed outstanding service under conditions which were at times exceedingly trying.

GENERAL E.F. BULLENE'S VIEWPOINT FROM SAN JOSE ISLAND

> Some of the people in Washington were quite irked for the reason that I saw fit to make haste slowly. However, in spending as much time as

> we did on water-borne sewage, constructing hard surfaced areas around messes, and putting up what was eventually satisfactory housing, I feel very proud of the fact that none of you gentlemen who participated contracted any of the many diseases which are so prevalent in the Tropics, and which, from my past experience in the Tropics, I felt sure some would if a decent base was not first constructed from which to work ... I met the sole survivor of the Gorrell party which, as you may recall, under Dr. Gorrell of London, conducted similar experiments in a jungle-covered island near Australia. They were a bunch of enthusiasts who went at it the other way and started carrying out experiments almost immediately upon their arrival with no thought of a sanitary base. Within four months [all but one] were invalided back to England and the United States. [Letter written by General Bullene to Professor Francis Blacet after the war].

Apparently, the troop ship named William Everts and its twenty-one members of NDRC Division 10 were caught in the middle of this argument.

San Jose Island, Panama

For a long period of time, the San Blas Indians had lived on San Jose Island, but it provided meager sustenance, and they had abandoned the island many years before. The army hired a group of San Blas Indians temporarily to return to the island and work as laborers and servants. The army provided housing for them, adjacent to the main base, Figure 5.1.B.

In preparing the station at San Jose Island, the army stripped the trees and heavy vegetation from a large hill that sloped up from the Pacific Ocean. The army installed roads, tent platforms and tents, dining halls, officers' club, medical facilities, storage houses, electrical power, clean water system, and sanitary flushing toilets. The officers' quarters were on the top of the hill, an example of which is given by Figure 5.2.A. Officers and the NDRC group lived in the same type of tent. Enlisted men lived in tents built close to each other and on the side of the hill. A friendly blimp appeared off the island one day, and Bill Shand rushed out to take a picture, which included soldiers' quarters in the foreground and an army warehouse in the background, Figure 5.2.B.

The tents were mounted on solid wooden decks well above the ground, were roomy with provisions for good circulation of air during the day and closed by mosquito netting at night. John Otvos and Bob Mills were roommates in one of these tents, Figure 5.3.A. John reported that scorpions

Figure 5.2.A. Avenida Central, officers' living quarters, top of the hill.

Figure 5.2.B. Blimp over San Jose Island, Panama, on cloudy, rainy day, 1944. Army warehouse in background. Tents for enlisted men in foreground.

Photographs by Bill Shand. Provided by Mrs. Rene Mills.

Figure 5.3.A. NDRC living quarters. Bob Mills and John Otvos, tent mates.

Figure 5.3.B. NDRC off to work. Bob Brinton and Dave Volman. 1944.

Photographs by Bill Shand. Provided by Mrs. Rene Mills.

loved the toilets, and special precautions were needed at each usage. In Figure 5.3.B. Bob Brinton was driving to work with Dave Volman in a four-wheel-drive weapons carrier.

THE JOB

Professor Francis Blacet, UCLA, was Head of the NDRC group, Dr. Bill Gwinn, UC Berkeley, was second in charge of NDRC. Professor Roscoe Dickinson was a working guest at the Panama test site. He made two trips to the island and stayed for extended periods of time, working with the Berkeley group to set up their micro-meteorological stations on his first trip and writing up the final reports on his second trip. These three principals of the NDRC group are pictured in Figure 5.4. It is interesting to note the shoes worn by these three leaders: Professor Dickinson wore clean street shoes, showing that his recent work had been in the office; Professor Blacet wore clean shiny army boots, appropriate for the boss who worked in the office and did field supervision riding in his Jeep; and Dr. Bill Gwinn wore the dirty muddy boots of a worker in the field. The full NDRC group posed for the photograph of Figure 5.5.

On June 14, 1944 Professor F.E. Blacet submitted a special report, a copy of which was sent to Edgewood Arsenal, that described the nature of the work being done by Division 10 of NDRC. I obtained a copy of this from Dugway in 1997.

> The responsibilities of Division 10 in the Project.
>
> The activities and responsibilities of Division 10 and its group of twenty-two men in the Special Project are summarized in the following four paragraphs.
>
> 1. The Division has undertaken to supply and maintain the non-persistent agent sampling equipment. To mention only a few of the most important items, it has supplied over one hundred recording milliammeters of the Esterline-Angus and General Electric types, 46 Dickinson meters, 48 hot wire analyzers and 100 vibrator pump-type samplers. The single job of keeping these instruments in good working condition under the handicaps to be met in the tropical forest is great, and there is plenty of evidence accumulating which indicates that the useful lives of these instruments, as now designed, will be short. The ultimate solution to this problem of deterioration probably will be to enclose all electrical circuits in dry air tight containers. This has been done for the recording meters but to do so for the existing Dickinson meters and other field units is not easy with the facilities at hand.

Professor Roscoe Dickinson
Professor Francis Blacet, UCLA, Head of NDRC group.
Dr. Bill Gwinn, UC Berkeley, second in charge of NDRC.

Figure 5.4. NDRC principals, San Jose Island, 1944.

Picture by Army photographer, given to Bob Mills, and given to H.S. Johnston

2. The burden of getting reliable results in the field with the non-persistent sampling equipment is carried largely by NDRC personnel. They have been very fortunate in having an able group of enlisted men to help in operating the field instruments. However, much of the credit for the splendid work performed by these men should go to the officers in charge and our greatest need is for more officers of the same caliber to share with the NDRC, and eventually take over the responsibilities associated with the field tests.

3. Almost without exception the micro-meteorological equipment used on the Project has been designed and furnished by Division 10. Four men from the Division have been busy working with the Meteorological

Back row:
George Doyle, George Cleland, Jim Pitts, Ted Gilman,
John Thomas, Lewis McCarty, Bill Roake
Middle row:
Bob Brinton, Bill Shand, Bob Mills, Phil Hayward,
Pat O'Conner, Clive Countryman, Chet O'Konski, J.M. Thomas
Front Row:
R.J. Grabenstetter, Dave Volman, Roscoe Dickinson, Francis Blacet,
Bill Gwinn, Jack Roof, John Otvos, Mike Kraus

Figure 5.5. Complete group of NDRC Division 10 on San Jose Island, August 1944.

Picture taken by army photographer, given to Bob Mills, and given to H.S. Johnston.

Section of the Project installing and maintaining instruments and collecting field data. In addition to acting as a service group for persistent and non-persistent tests, the Meteorological Section is endeavoring to make a complete survey of the micro-meteorological conditions which prevail on the island. Dr. Roscoe Dickinson has been particularly helpful in this regard, and it is hoped that he will be able to continue devoting much of his time to the Project.

4. To a large extent the Commanding General and the Technical Director of the Project have entrusted the planning, execution and interpretation of non-persistent field tests to Division 10. We appreciate this evidence of confidence and are trying our best not to disappoint these men.

The tests began by a series of static ground-level bomb bursts. For elementary education on how gases would move in the jungle, the army tested first a large bomb filled with butane and a few percent of nitrogen dioxide. Butane is non-toxic, and red colored nitrogen dioxide was a visible tracer. The group from the University of California had successfully fired several smaller butane bombs. The bomb was placed on the ground and detonated, but it caught fire and burned up. A large bomb filled with hydrogen cyanide also ignited and burned. They thought they had discovered something new, although we had seen the same thing in Florida in November 1943. Dickinson (Chapter 4) had suggested the cure, which was to dope the hydrogen cyanide with a few percent of gasoline, and in the few Florida tests of this idea this prescription worked. NDRC members placed their samplers for chemical analyses and made micro-meteorological measurements, and the army provided a circle of goats. Phosgene bombs did not ignite, but phosgene reacted strongly with jungle foliage effectively removing a significant fraction of the poison. Cyanogen chloride did not ignite; it too reacted with foliage but less strongly than phosgene. Because of this loss of vegetation, they could use an area for only a few tests. The need to prepare new test sites added to the work of soldiers and NDRC workers. The NDRC crews had to do some bush whacking to locate their samplers on a grid and to give access to them.

Most tests involved bombs dropped from aircraft. As the first bombing test, a B24 formation delivered a pattern bombing to the target area, but this test failed because of some mistake made by the Chemical Corps. John Thomas told me that the bombs went into the ground without exploding. Eventually they completed many successful tests and reached significant conclusions (W.A. Noyes, Jr., page 323).

Below are quotations from the June 14, 1944 special report submitted by Professor F.E. Blacet to Edgewood Arsenal, which discussed earlier tests. I obtained a copy of this from Dugway in 1997.

This gas [cyanogen chloride] is today thought to be the most promising nonpersistent agent. It is poorly stopped by all canisters except those containing ASC-impregnated charcoal. This is particularly true for humidified adsorbents. Equilibrated German or Japanese canisters will provide protection against moderate concentrations for only a matter of minutes.

Static and Aerial Tests with 8-1000 lb Bombs. Two tests have been performed with CC [cyanogen chloride], one in which the eight bombs were placed in a rectangular pattern approximately 200 ft apart and fired statically. In the other, the bombs were dropped from 3000 ft, landing in a pattern somewhat different than that used for the static test but probably comparable to it without serious error.

a. The wind velocities were practically zero and in both tests the gas cloud moved very slowly down the little valleys and stream beds. The cloud remained visible for one and three-quarters hours in the case of the static test.

b. ... The cloud remained visible for at least one and one-fourth hours. One hour after release this cloud was seen to reach an unmasked goat with a concentration sufficiently great to kill it in approximately two minutes.

c. Sufficient gas remained at the point of origin in both tests so that masks were necessary for at least one and one-half hours.

The AC [hydrogen cyanide] and CC [cyanogen chloride] protection of U.S. canisters is considerably higher than that of any other canisters.

THE BERKELEY GROUP

Bill Gwinn, Stan Winkleman, Clyde Countryman, and John Thomas represented the Berkeley project. They measured wind speed by a hot-wire anemometer placed in open air, and they measured phosgene or cyanogen chloride concentration by combustion on a hot wire contained in an aspirated cylinder. In each case a current of electricity passed through a platinum wire to bring it to high temperature. An increase in wind speed slightly decreased the temperature of the open-air hot wire, and an increase of phosgene increased the wire temperature of the enclosed hot wire as the gas burned up. Each grid site included two hot-wire instruments, and many sites were required to trace the motion of a gas cloud. It required a large amount of electricity to run their equipment. The army shipped their heavy electrical generator from Florida.

During a test, the generator was located well away from the target zone, and heavy wires went from the generator to the test site and along the grid of samplers. The generator put out 120 volts alternating current. When all the stations were connected and in operation, the voltage in the line dropped to such a degree that many of the stations could not operate. Bill Gwinn solved this problem by using a large transformer to boost the electricity to four hundred volts and used small transformers at each station to reduce it to 120 volts. In this way they were able to operate all of their stations. John Otvos recalls that he saw Bill Gwinn splicing together two wires each at four hundred volts while Gwinn stood with wet boots in a puddle of rain water.

NORTHWESTERN UNIVERSITY GROUP

There was a large group from Northwestern University, and they operated the one hundred vibrator pump-type samplers and other samplers. I have no record of their detailed activities.

CALTECH GROUP

> Dr. Dickinson spent several weeks at San Jose on two different occasions. His contributions to the meteorological program were very substantial. [W.A. Noyes, Jr., page 322].

John Otvos, Bob Mills, Bill Shand (skilled and daring mountain climber), Mike Kraus (dedicated and advanced bird watcher), Ted Gilman, George Cleland, and Phil Hayward were the NDRC group from Caltech. Among other things, they operated the fifty Dickinson meters or "Egberts." For each test they had to place them and the recorders in the jungle on the sampling grid. They turned on the water flow and started the recorders at a fixed short time before the bombs were dropped and withdrew from the test area. As soon as the bombs were dropped, they put on their gas masks and rushed to the sampling site to check and maintain the operation of the Egberts. Afterwards they labeled and removed the recorded data and removed the equipment to a water-tight storage van.

As Bill Shand was setting up equipment before an air-drop test, he was in communication with the field leader about where to go next by way of a "walky-talky." In the middle of an instruction, Bill said, "Wait a minute. There is a boa constrictor in the path." The phone went silent for what seemed like a long time, and the field leader tried to reestablish the connection. Finally Bill came back on the line, saying, "It got away."

* * *

Figure 5.6.A. Major Tinkler, Captain Quate, Bill Shand with NDRC, at work together.

Figure 5.6.B. Bob Brinton in a mangrove swamp. San Jose Island, Panama, 1944.

Photographs by Bill Shand. Provided by Mrs. Rene Mills.

The NDRC people and army officers worked together, as illustrated by Figure 5.6.A. The Major, the Captain and Bill shand of NDRC were similarly dressed in work clothes and were all doing the same job. Bob Brinton was photographed in an almost impenetrable mangrove swamp in Figure 5.6.B.

Figure 5.7.A. Recreation, NDRC: Baseball between ocean and jungle.

Figure 5.7.B. Sand sculptors. What were these guys thinking about? George Cleland, John Otvos, Ted Gilman, Bob Mills.

Photographs by Bill Shand. Provided by Mrs. Rene Mills.

There was some recreation for workers on the island such as fishing, swimming in the ocean, playing baseball between the ocean and the jungle, Figure 5.7.A, and creating a sand sculpture of a lovely female on the beach,

Figure 5.7.B, which they named "Sandy." Figure 5.7.A gives some perspective of the great height of the jungle trees.

* * *

Through their letters back to the states, John Otvos and Bob Mills supplied information about how the NDRC personnel interacted with the environment and with their job. I give a separate section to the contributions of each of these, in which I quote selected portions of their letters.

LETTERS FROM JOHN OTVOS

> **April 9**, Easter. Dear Mother and Dad, A batch of letters came from you yesterday ... The food is a little better, or did I never tell you it was no good? At least now we have jam on the table always and every day we get ice tea.
>
> I have to do some teaching [to the officers about how to operate some NDRC equipment]. Yesterday I taught and talked for five hours, and it was not easy, believe me. It's not like at Caltech. Here every man knows a different amount by a long shot, so no matter what you do, you go too fast for some and too slow for others.
>
> Mills is still in the hospital with his infected foot ... He's a very important guy, as things stand now so his illness is inconvenience to everybody ...
>
> Got a letter from Witkin who is still OK in Italy, but he says he's been through a lot of fighting already. I'm sort glad we have no radio to tell us how many planes we lost over Italy and Romania each day ...
>
> I'm feeling fine, but scratching myself all over. The sand fleas have used up almost the available space on me. Pretty soon they'll come along and say—this is where I came in—and then go away. Love John.

Army and NDRC people could occasionally take off weekends for recreation on the mainland. A shuttle boat, operating from the recently erected pier, carried mail and a limited number of passengers to and from the mainland.

> **April 17, 1944**. It was really a thrill to hear your voices via short wave radio the other night ... Saturday morning I was in Panama City after spending my first night at the Tivoli Hotel, which is quite picturesque. At 11 AM I went to the Radio Company and made my reservation for 8 PM from Cristobol. I knew we would be on the Atlantic side that night. We went over mainly to see some of the canal from the train and to eat and drink at the Washington in Colon ...

If you are clever about it, it's possible to swim in the Pacific and then two hours later swim in the Atlantic because the train takes only 1 1/2 hours. Well all we did on Saturday was swim in the Atlantic at the beach of the Washington Hotel. Then cocktails at 5 on the porch overlooking the ocean, followed by a fine dinner. After the phone call, we went to a night club and saw the worst floor show ever presented to me. By 12 we were back at the Tivoli Hotel in Panama City ...

Now for your questions:

1. We have no radio. Haven't heard a program or news since we left USA.

2. I have no idea how long I am staying and if I did I shouldn't say.

3. I am very much up to par and feel swell except for a few itches I told you about.

4. I sleep plenty, 10 to 5 every night.

5. Women? No, Madam. And I can't imagine what made you put in that question right after that question about do I sleep enough. I have danced a couple of times since I left but for the most part all we have a chance to do is to talk about women, and that takes up lots of time.

6. Homesick? Not yet. I'm too busy and the trip is still too much of a novelty.

7. Your mail has not been censored yet.

8. Never mind the package. It would take too long and if it's food I would not get much of it as you know.

9. I stand the climate very well, even though it requires at least one shower a day.

10. The food is getting better and we have enough. We sit at tables and it is served on a tray with compartments. The whole meal at once.

11. We eat with the officers—army food, and we pay for it, 25¢ a meal. We also pay $6 a month for valet service. A San Blas Indian makes our beds and shines our shoes.

12. Laundry done by the army, cheap and fast—one week...

13. We wear mostly army clothes but it is all optional. Here at camp it's mostly work clothes, except clean khakis at dinner. When I go to civilization I wear civilian clothes...

14. The beach is fine but there is no shark net yet so we aren't allowed to swim. Soon we hope we will.

15. Holidays will be rare, I guess. That's why I took the opportunity last weekend.

April 24, 1944. Lately I've found your letters are coming more slowly ... Tonight I was visiting Mills when I heard the radio they have in the hospital. There was a Fred Allen show being played ... Well, soon the Fred show began to sound familiar and I realized it was one I had heard in the USA in New Orleans about 10 weeks ago. I am now disillusioned about short wave rebroadcasts.

May 5, 1944. The boss arrived yesterday, but I have not seen much of him yet. His humor is just as dry as ever though. It's late and I'm sleepy.

June 4, 1944. My regular letter writing schedule has been temporarily interrupted due to a trip to Panama with Roscoe G. and Bob and Ted. We are more or less acting as guides for our boss through the Republic, but after a long spell without going to the city, it is all dull to see the same places over and over again ... The shops on the main streets are either filled with souvenirs or contain ordinary clothes and things which you can buy anywhere ... The night spots have a few Spanish singers but there isn't a bit of Panamanian to them. The shows aren't the slightest bit sexy ... Here again I am forced to a comparison with Puerto Rico, where a little restaurant in San Juan had a little orchestra and one good singer. The Puerto Ricans themselves came there, not just soldiers and sailors and it gave you a feeling of being in a quaint and different land.

June 12, 1944. I feel happier right now than at any time since I left home. The satisfaction I get out of work is unchanged, but the news from home today is great ... Don't worry about my health. We take all the necessary precautions against mosquitos, liking sleeping under nets, and not going around after dark with any uncovered skin unless it is protected with insect repellent. Besides, there are surprisingly few mosquitos.

June 23. Tonight two newly made Captains passed out cigars and I just finished mine. It reminded me a lot of home—the odor and the burning sensation in the eyes ... I had my hair cut while finishing the cigars too, and all in all its been a pretty good day. Good food, a letter from Allan in which he tells me of the excitement at the studio on D day, and a new issue of Time ... Just figured out my per diem. To date I have $756.42 coming. Of that I have received $500 advance, so $206 is still on the right side of the books. Feeling fine—but there is an awful smell here. I better get rid of the dead mouse.

June 27. Nothing much has changed around here except that I have a greatly increased feeling of restlessness and frustration. I want to get through and come back. Most of the others are like that too. And I wouldn't be surprised if when I did get back, I changed jobs. It's all indefinite in my mind, but there are some pretty interesting spots open at Cal Tech which I might take advantage of. This is all between us. On reading this it seems that perhaps I'm implying I had differences with the boss. This is emphatically not the case. In fact we talked about it while he was here. And it is not a personal matter involving anyone else, either. Everyone is still just as cooperative as ever and we get along fine. But after all, I've been doing the same sort of thing for three and a half years now and a little change would be welcome. Besides I don't think I'm doing as much as I might be.

LETTERS FROM BOB MILLS

March 31 and April 7, 1944. Friday night. Left Panama last Tuesday & have reached final destination. In just 7 more days I will have been gone two months. I am hoping now to be back home by labor day.

John & I share a tent [Figure 5.3.A] & immediately opposite are Shand & Gilman, neighbored by Hayward and Cleland—so all Caltechers are sticking pretty close ... We eat breakfast between 0530 and 0600, eat lunch at 1130, followed by siesta till 1500 & then dinner at 1900. All in all its a full hot dusty day.

On Sunday I contracted a foot infection that had me confined to my tent for several days. Since the ailment failed to get any better, the medics finally moved me to the hospital where I am now & have been for 2 days.

I just have to let you know how much I liked your last 2 'chock full of news' letters. It's getting so I'm the best informed man here about doings back Caltech way.

Excoose me while I shift—its uncomfortable with my foot sticking up in the air like a dead blue jay.

April 8. [I slightly rearranged these questions and answers]. Saturday night. These are the answers to the questions you threw at me at the end of the letter I received from you today:

1. Can you go swimming?	No, not yet.
2. Are you working hard?	No, not now.
3. Are you playing any bridge?	No.
4. Can you speak espanol?	Non fluently.
5. Do you get to civilization?	Not so far.
6. How are all the boys?	Fine & healthy.
7. Are there any lovely native women?	No.
8. Are you wearing your hair short?	Yes.
9. Is it hot?	Yes, as hell.
10. When are you coming back?	Dunno.
11. Can I come down to see you?	Try it.
12. Do you have an extra shoe coupon?	No!!

> Wish you could see the homey little tent John the padre & I share. Still I believe I prefer Caltech with its inside plumbing and little things like that there.

Bob was out of the hospital according to his letter of April 17 and back in the hospital April 24. All of his shoes were sterilized by autoclave, "melted them right down to the raw materials." He was finally released from the hospital on April 27 and had difficulty in getting replacement shoes.

> **April 29.** Saturday night. We had some excitement yesterday when T.S. Gilman turned over the truck he was driving. He was able to abandon ship & escape before the car rolled on him.
>
> Tomorrow morning at 10 there is a softball game scheduled between the NDRC and the army officers ... The ball field is located on the side of a hill and from home plate only 2 outfielders are visible.
>
> Tomorrow for the first time there will be a church service here— so the bulletin board reads. About the doxology I'll think of you sitting in the Pasadena Methodist Church.
>
> Also, on the docket for tomorrow is game hunting by the officers.
>
> **May 1.** Monday night. Although Sunday was a day of rest, all the NDRC fellows worked all afternoon.
>
> They now have beer on stock at the PX here. It is Panamanian 3.2% brew.
>
> **May 5.** Friday night. R.G. arrived ... Dickinson is his same old jovial self and after two beers he was just 3.2% more jovial.
>
> And now for a short summary of the vertical flood we had last night. There was a time there when I thought John and I would have to put oarlocks on out tent and row it back uphill. Fortunately there were no serious drownings.
>
> You asked about my foot. It's all OK and I have neither gained or lost any toes.
>
> **May 11.** Thursday noon. Incidentally some of the horny guys here have voted Ann Miller as 'the girl they would most like to pick ticks off of.' I guess the fellows are tick happy for we're plagued by them constantly. In case you've never had encounters with any of them they look like this [small drawing] and burrow into your skin and when you try to pull them off, the head usually comes off and festers.
>
> **May 20.** Sat. afternoon. I've been working night and day & in between. My morale, which had been dragging in the sand, reached an all time high when your 7-pg letter arrived. Thanx again for making me the best informed about Caltech. Naturally I pass on choice tidbits to my pals, and whenever the news gets out that Mills got a letter from Rene, well I'm swamped with inquiries.

I spent a day working at the beach with R.G. and got the end of my nose sunburned. Really, Rene, you should see this beach—fine white sand bordered by coconut palms.

I know you would have enjoyed being on the deep-sea fishing cruise I took with Ted & Mike last Wednesday. Incidentally we 3 caught the only fish—Mike a mackerel and Ted & I each a 10 lb bonita which we were barely able to battle on board.

May 23. Thursday night. Next weekend Ted, John, & I are going to try to make it to the big city for a joint celebration of birthdays—Ted's 26th and my 22nd.

Even now I hear the light pitter-patter of rain as it strikes our tent roof by the bucketfuls. Last night I was rudely awakened by rain in the face. It seems that John was sleeping through it (or pretended to) so I had to close the tent flaps. Naturally I went right back to sleep and thought no more about it. This morn when I stepped lightly from the tent, I narrowly missed my doom. If my eyes had been closed 1/16 inch further, I would have strolled over a 30 foot (9 m) cliff. It seems the 'light rain' of the night before had washed away our hill so our tent was perched precariously on the newly formed precipice. We expect to literally 'go over the hill' tonight for every now and then I hear the tent frame groan as more and more dirt is washed from under us. I've always wondered what the sensation was to go over Niagara Falls in a barrel. (I'll include a resume of same in next letter).

June 1. Thursday night. I hear rain on the tent now which is nothing new. Its been with us for weeks. Our tent hasn't washed away yet—but soon. I'm almost accustomed to waking up wet, working all day wet, and then retiring wet. If it weren't for the interesting and extensive work, I might notice it.

Tomorrow is Friday and John, Ted, R.G. and I have reservations for a trip to Panama—mainly for celebration purposes and to show R.G. the sights.

The rainy season has brought with it mosquitoes and that means mosquito nets.

During Roscoe Dickinson's trip to San Jose Island, he was escorted to Panama by his Caltech crew and shown the sights. As they were doing so, Professor Dickinson, to his surprise, encountered Commander Robert Dickinson, his son, who was also sightseeing in Panama and on his way to elsewhere. Neither realized that the other was anywhere near Panama.

June 15. Thursday night. The rain of which I spake all last month let up for a while and living was again pleasant.

Remember the light which used to burn in 65 Crellin, well it shall shine again soon for it won't be long before R.G. will be there. See if you can note any changes in him, if he seems slightly uncivilized, crude, horny, and so forth. This life does that to one you know. Really though he acted as if he actually enjoyed his stay here.

June 23. Friday night. I'm in a rare bad mood tonight—so excuse it pleez. It seems I was out working late tonight & when I got to chow there were no pork chops left—only a cold slab of salami. Naturally John was not similarly affected for he's always standing outside the mess hall waiting and usually is the first in line when they open the door. He was particularly obnoxious tonight for he sat here in the tent getting a haircut and smoking some foul cigar which a captain was passing out to celebrate a promotion. It was free so he took it. Yes, I had one too but it went out soon after I lit it. So of course I threw it out.

Bob's letter continued: He [Mike] saw R.G. off on the plane yesterday. I sure miss him for he's a regular guy. Used to buy me beers at the PX all the time. The PX opens at 4:30 every day and Dickinson was always first in line (Just like John in the chow line).

You should see our new tent. We really have it furnished comfortably with a light switch which hangs down between the beds so we can turn the light off after tucking ourselves under the mosquito netting. Atabrine also seems great stuff to combat malaria.

July "0". "Saturday night. Bill Shand became allergic to atabrine, was hospitalized, and switched to quinine."

July 6. The censor cut one line from this letter, which was the only such cut in all of Bob's letters.

July 7. Friday night. "Oh yes, I meant to tell you about last Sunday. It was an official rest day, so John, Ted, Bill & I got a brand new softball from Special Services & started through the jungle to the beach. It's a nice long sandy beach, fringed with coconut palms & just the place for some outfield practice. [Figure 5.7.A.] After about an hour of knocking out flies (only 2 went in the Pacific), we had our usual surf bath (the water is about 83 °F all the time (28 Celsius)) & then retired for some sunbathing. Ted, reverting back to his childhood, started digging a hole in the sand. This prompted me to model a face in the wet sand—and this prompted El Lobo (Shand) the idea of modeling a lovely reclining damsel. Naturally we all pitched in gladly to help him—each to his own speciality. The results were very pleasing—only our demure girl turned out to be a creature approx. 6'6". We decided to call her 'Sandy,' and as the sun got low over the ocean and cast revealing shadows all over the place, Shand was out snapping pictures like madly. [Figure 5.7.A, B.]

July 11. Tuesday night. The big news around here has to do with camp improvements. Tonight we initiated a new tent theatre just erected. The General addressed the masses and did a fine job of moral-lifting (anyone whose moral was not lifted was court-martialed, immediately). The improvements he mentioned were as follows:

1st the erection of a shark net
2nd opening a library
3rd construction of a baseball field
4th etc. etc.

July 12. Bill Shand came in and started banging away on the tripewriter and spitting out bright chatter. Incidentally, I think he's getting jungle jollier (or is it jungler jolly?) (or is it jungler jollier?) every day. I've been working with him the last couple o' days & today he came within a hair of clipping off my left ear [with a machete]. He said he thought the vine would be harder to cut than it was, but I think it was a sloppy attempt on my life. Could he be jealous? Rene, dear, I think you can now add Shand's name to your list of hungry pursuers.

John ... tried killing them [cockroaches] with flit, but this tropical variety just stands up on its hind legs, breathes deeply of the stuff, beats on its chest, and then goes about its business of crudding up the suitcases with renewed vigor.

It's raining like bloody hell today & naturally I was saturated.

I ate with Mike and Phil at the new officers' mess tonight & the food is excellent. We have new Chinese cooks, and the San Blas Indians wait table. also there are plates to eat from—no more tin plates.

August 2. Wednesday afternoon. The last bit of leave I got was last June 2nd when R.G. was here when we celebrated birthdays. Well that's been two months ago—and believe me we need some city life again. Anyhow Ted, John, and I cooked up this little vacation. We were able to squeeze out a 7 day pass which started yesterday with the boat cruise to Panama.

Last night we celebrated at the Tivoli Hotel with appropriate drinks & dinner. We finished just in time to catch the last show.

The new [baseball] field is perfect—and the set-up in general furnishes some incentive for staying—but not enough. [Figure 5.7.A.]

August 7. R.G. Dickinson returned to Panama. Bob and all returned to the island.

August 12. Saturday night. [Swimming in the Pacific here] would be tolerable if the sharks, barracudi, sting rays, mantas, etc. were filtered out. [The general was not correct in July to say the shark net was in place.]

Ted had written a letter to R.G. asking him to bring down certain apparatus when he came down the second time. On his arrival, Ted asked him about the items & R.G. replied that Ted's letter had reached him only 15 minutes before he had to catch the plane. Ted's face got long and he mumbled something about it being too bad he couldn't get the stuff. Dickinson, with typical Dickinsonian humor, said, "Whata you mean? I had 15 minutes, didn't I." And he handed him the stuff.

Which leads up to the second point. You mentioned about the dean of the graduate division [R.G.D.] looking so sad & lonesome in his aimless wanderings through Caltech's stately halls—well he's certainly in his element here and all set to hit Panama City again.

August 16. To get back to my predominant mood, I might mention that I think of you loads, miss you lots, and can still conjure up an image of you taken from a certain Sunday Feb. 6, 1944 at the Alhambra depot. It seems odd that I should feel that I know you so well even though I've been pestering you 1 month out of 7 or 8. At least I can still pester you by mail though not nearly so much fun as in Harold's convertible.

Speaking of H. Sledge Johnston reminds me that I haven't heard from him since I left the states. R.G. was able to peddle some meager information, but perhaps you have access to more.

Work here is ready to hit high gear pronto—and this means that Departure day will be that much sooner.

FROM RENE SCOTT TO BOB MILLS

March 28. Got your letter today written from Panama. I guess it is safe to write that since you got away with it. As yet nothing has been cut out. The sightseeing sounds marvelous, and I only wish I were there enjoying it too.

Dr. Yost has been coming over to the lab quite regularly. People were talking about how bad he looked, but to me he looks wonderful compared to the last time I saw him in the hospital. His book on Inorganic Chemistry will be out in April, and he's pretty happy about that. I guess Horace [Russell] will be rather proud, too. He's co-author, you know.

There's one question you asked that I'd like to answer; namely about how a certain party of Corey's project is getting along. Well frankly, old boy, as well as could be expected, if you don't expect much. It still arises at 7:00 or shortly thereafter, works about 8 hrs., and comes home to a dull evening at home.

Rene was also working on a secret war project, which concerned chemical analysis of rocket propellants and their added stabilizers. Rene was an expert in analyzing captured enemy rocket propellants. The project developed chemical stabilizers for the rocket fuel and methods for rapidly artificially aging them. The project had a large volume of work, and it had expanded to include a large number of Caltech chemistry graduate students. Bob complimented Rene on her sending news and gossip from Pasadena, to his delight and to that of the others from Caltech. For example, here are a few quotations from one of her letters:

> **May 7.** Sunday nite. There's lots of news of one sort and another. For instance, did you know that Bob Boykin got married today? The girl is from Beverly Hills. She's tall, good looking and very swell.
>
> Weddings and showers are big items these days. Fran and Evie are having a shower for Marydell on Friday, and Mary Sease and Hig are doing the same for Maryjane on Saturday.
>
> Judy got back today. She went as far as Salt Lake with Betsy & Marge Laird and then spent a week in San Francisco and Palo Alto with Ginny. Betsy and Marge went on to Evanston in the station wagon and from there I guess they go on to Washington. Betsy wants to get into Red Cross overseas service & intends to do some inquiring while she's in the old capital.
>
> Spring fever has hit Crellin like a seven man football line. Besides those about to be really tackled, we have young couples wandering hand in hand all over the place. Kent Wilson and Gladys Acevedo are seen together most frequently. Fran Colt has been known to have been seen with Austin. Evie Bradford and Heintz are regulars, but Heintz leaves in a couple of weeks for the armed forces. M.E. and John are out quite often. Al Senear is victim of the latest rumor of romance. Golding says he (Al, I mean, not Dave) is going to be married, but then Golding is a cynic about such stuff and can't always be regarded as reliable. Al Soldate is getting married in June (A girl from home).

Rene's news was not all about young lovers at Caltech. Some of her news was serious and tragic:

> **August 6.** Part of my awful mood is due to news Dr. Yost told me today. He came into the lab looking for some exciting news to cheer him up. He'd just gotten a letter from Dick Dodson saying that Horace Russell had been thrown from a horse & wasn't expected to live. He had a bad brain concussion, but no fracture. A specialist was flown in, but he said an operation would not help. There seems to be almost no hope. I shall keep you posted if more news is available.

The next day the news came that Horace had died. Horace and Rene worked on the same war project; he had recently completed his Ph.D. Degree with Yost and was co-author of the new book by Yost and Russell. The accident did not occur as a part of his war work.

* * *

When I left Pasadena in November 1943, I left my 1937 Buick convertible sedan in care of my friend Bill Lipscomb. Bill let Bob Mills have use of the car during January 1944. In her letters Rene occasionally mentioned the car or the indirect effects of the car, and I quote a few examples:

> **April 29.** Bill Lipscomb heard from Hal today. Will let you know any new news when Bill reports. Speaking of Bill reminds me you will miss his wedding. I guess it will be in the last of May.
>
> **May 29.** It's funny how that one short month of January remains so clear.
>
> **June 18.** Because it is such a beautiful night, I think I'll be reminded of another one—when we drove to Mt. Wilson.
>
> **August 6.** Tomorrow a gang of us are going back to Shangri La for a swim. Someone is taking Hal's car—and that is half of my trouble. I fully intend to burst into tears when the car drives up and someone else gets out of it. It will be a brave attempt on my part to keep from looking for you. Among those going is a new girl on our project, Alice Shannon, whom T. 'Wolf' Waugh has his eyes on—so Mary Evelyn informs me.
>
> **August 17.** Thursday night Reuben Wood's folks had a crowd out to a really superior dinner. Mrs. S. (can't spell it), Al, Norm, Art, Fran, Ken, Evie, Dick, Dave Schoemaker and I <u>all</u> managed to squeeze into Hal's car... It was an odd sensation to see Al drive it and even odder to sit in the back seat.
>
> Skip to Sunday—nothing much of interest Fri. & Sat. Hal's car again used to transport people to Shangri La for a swim—Al, Tom, Norm, Andy, Ken, Alice, M.E., and [Rene].

FROM JOHN THOMAS

Undoubtedly advised of the opportunity, a Smithsonian biologist was in residence on San Jose Island. He collected immense numbers of animals, birds, and fish killed by gas. Because the island was far offshore from mainland, many species were different and had not been identified before.

John Thomas had an interesting experience wading down a stream several miles to get to a beautiful beach, dodging crocodiles in the stream and sharks in the surf.

Special tests. We observed several special tests the army carried out. One was the use a proximity fuse, only recently developed, to detonate a nitrogen dioxide bomb over a barge floating offshore. The pilot's aim was perfect, as was the fuse and the red cloud enveloped the barge.

Before we finished our work a persistent gas group moved in, tested mustard gas, etc. and protective clothing.

MAJOR CYANOGEN CHLORIDE TEST[#]

> In August and September 1944 several attacks were performed under operational conditions with the aid of the 6th Air Force on a scale which, as far as we are aware, was never equaled in gas experiments during this war. The data obtained were more complete than those ever obtained before. [Noyes, pages 322–323].

Simultaneously from several photographic stations, the Chemical Warfare Unit recorded moving pictures of a major test using cyanogen chloride as the agent. A fleet of B-24 bombers dropped a large number of bombs on a selected portion of the jungle (Figure 5.8.A), and soldiers fired mortar shells filled with cyanogen chloride into the same area (Figure 5.8.B). These bombs and shells produced a conspicuous large cloud of cyanogen chloride above the canopy and within the jungle. As stated in Chapter 4, leaves in the canopy absorbed sunlight during the day and became warmer than the air above, so that above the canopy there was an unstable lapse rate, and the gas billowed upward (Figure 5.9.A).

The leaves in the jungle canopy were so dense that they absorbed essentially all the sunlight, even at mid day. Sun-heated air in the canopy was warmer than air near ground, which produced a temperature inversion inside the jungle. The gas cloud in the jungle at midday, was flat at the top, showed little vertical mixing, and crept through the jungle, maintaining high concentrations. The appearance of the gas cloud was similar to that of smoke clouds on clear nights at Rosamond Dry Lake, when there were large temperature inversions. Figure 5.9.B. Three goats were tethered in the

[#]On November 8, 1998 the television news program, CBS 60 Minutes, showed a small fraction of the army's moving picture record of the August test. John Thomas saw this program and ordered the video of this news program, and he gave it to me, knowing that I was working on this book. I ran this tape repeatedly and noted the parts of it that were copied from the army film. With a 35 mm still camera, I took numerous snapshots of a television screen as I ran through the videotape. Below I give a summary of what the army pictures show.

Figure 5.8.A. Cyanogen chloride bombs dropped from aircraft.

Figure 5.8.B Cyanogen chloride poison gas delivered by mortar shells, Panama, 1944.

Photographs by U.S. Army. Pictures obtained courtesy Archives CBS 60 Minutes.

Figure 5.9.A. Cloud of cyanogen chloride above Panama jungle, 1944.

Figure 5.9.B. The cloud of cyanogen chloride in Panama jungle approaches three goats, one with U.S. gas mask, one with Japanese gas mask (lying down), and one unprotected.

Photographs by U.S. Army. Pictures obtained courtesy Archives CBS 60 Minutes.

jungle in the path of the approaching cloud of gas. One was fitted with an American gas mask, one was fitted with a Japanese gas mask, and one had no protection. Secret laboratory tests by NDRC showed that cyanogen chloride fairly rapidly penetrated captured Japanese gas masks.

When the gas cloud reached the three goats, all of them were highly distressed. They leaped, stumbled, and thrashed on the ground, but the rope around each neck held the goat close to its post. The goat with no gas mask and the goat with a Japanese gas mask died. The goat with the American gas mask survived.

The pictures of these goats show shocking cruelty to animals. The purpose of these tests was not to kill a few goats, but to learn how to kill large numbers of people. When we go to war, we turn our affairs over to an entirely different priesthood.

* * *

Bob Mills in his letters depicts the final work at Panama:

> **August 18.** For the past week, we have been living on C-rations—naturally I in particular take a dim view of such behavior—and the view becomes even dimmer when I realize that we have to live on C-rations again next week. Dickinson also suffers with us and gets off some choice remarks at meal times. [NDRC group photograph taken at this time, Figure 5.5.]
>
> **September 1.** Friday night. Seriously I'm really not as bad off as I might sound. It's just that a lot of the guys are leaving pronto for the states. Yes it's true, Otvos sails tomorrow for Panama where he will immediately wave a priority rating at Pan American Airways. His delay on the Isthmus, though, may be infinite, and in the event that he does not get passage right away for New York, then he plans to return directly to L.A. He figures (in true Otvos style) that if he can't get to N.Y. in time for his roomy's wedding, then it would not be exactly in good taste to interrupt him in the midst of honeymooning.
>
> Bill Shand and Phil are slated to leave in 2 days, but I guess their plans include a stop at Mexico City where Shand will again try to break his neck climbing mountains.
>
> **September 11.** Monday night. On the morning of the 9th, just before I left 'there' for the Isthmus, I got in a whole morning of deep sea fishing. Four of us trolled the Pacific by motor launch, lounging in deck chairs in the stern of the boat & equipped with the best ocean rods & reels the army's special service outfit could muster. By noon we had caught 65 pounds of fish ... 2 tunas and 2 mackerel. the largest mackerel (I caught naturally) was over 3 feet wheelbase.

> When El Padre left, my sailing date was set at Sept. 13 … Mike and R.G.'s priority called for departure on the 14th. But alas a beaker of cold water has been dashed in our respective faces and I sadly wake up to the realization that I am doomed to two weeks worth of stay on the Isthmus. The four of us are now staying at Albrook field near Balboa. The set-up is too good to enjoy alone. We live in a Colonel's apartment & the luxury and accommodations have yet to be equaled in my inexperienced life.

Their priorities for departure were canceled until they finished writing the reports. Dickinson played the major role in writing the final report.

September 20. Wednesday noon. Bob, Mike Kraus (the advanced bird watcher), and Dickinson were definitely scheduled to leave Saturday, September 23. Bob warns Rene what the NDRC crew will look like and do when they return:

> 1st, in the midst of conversations they may partially disrobe in the search for some annoying tick.
>
> 2nd, they may remove shirts, trousers, and what have you just to feel at home.
>
> 3rd, they may constantly scratch themselves in conspicuous as well as inconspicuous places.
>
> 4th, before donning shoes or wearing apparel, these will be carefully shaken to loosen & discourage intimate contact with scorpions, centipedes, and cockroaches.
>
> 5th, the language will naturally be obscene, unintelligible, or both. Translations however can be made by any G.I.
>
> 6th, All will have that slow 'Panama shuffle' and some may observe siesta (1200–1500).

* * *

John Thomas reported: "Upon leaving we turned over all the equipment, after sessions showing them how to use it, to Chemical Corps people. I trust it served them well."

Southwest Pacific

In the late summer of 1944, W.A. Noyes, Jr. in Washington invited NDRC workers in Bushnell and on San Jose Island, Panama, to volunteer to go to New Guinea to join a chemical warfare station there. Robert Brinton and Bill Shand were the only volunteers. They traveled to New Guinea with a group of army officers, and they moved from station to station, presenting recent NDRC achievements and methods to front-line officers. Bill Shand joined the Far Eastern Technical Unit in New Guinea, and from January to November 1945 he helped them carry out field trials.

On October 17, 1944, Professor Carl Nieman of Caltech sent Don Yost a hand-written letter from New Guinea, saying in part: "Any sort of travel, is rather rugged and extremely casual. I have been able to get about and have seen several spots of considerable interest. There is no doubt that the boys out here have had rough going and anything that can be done should be done regardless of red tape and sheer stupidity." [Courtesy of the Archives, California Institute of Technology].

After a time in New Guinea, Bill Shand was given, probably volunteered for, a special assignment. He was to fly as the passenger in an open cockpit of a fast Air Force fighter plane close above the beaches of Japan just after the beaches underwent heavy bombardment prior to invasion by our troops. Bill was to see if he could smell mustard gas. Detection by smell was much faster and more sensitive than any instrument then available. He received some practice by flying over beaches in remote New Guinea on which ground-based mustard bombs of various sizes had been detonated. The war ended before he carried out this assignment over Japan. After the war Bill told me the officers in New Guinea had evidence and believed that the Japanese planned to soak their beaches with mustard as invading troops approached.

CHAPTER 6

FLORIDA, 1945

OUR JOB

The Dugway Proving Ground Mobile Field Unit of the Chemical Warfare Service (CWS) in conjunction with National Defense Research Committee (NDRC) Division 9 carried out field tests, mostly using mustard gas, in central Florida from January 1944 to the end of the war in August 1945. I was in charge of the micro-meteorological department during this period, both for the field measurements and for transcribing the data to a form useful to the army. Members of NDRC placed clean chemical samplers at the site before a test, the samplers pulled in air and extracted mustard gas vapor over a measured period of time, and the apparatus was picked up and carried back to the laboratory in Bushnell for chemical analysis the following day. This operation involved hard tedious manual labor. We routinely took raincoats with us during the field tests and continued to work when it rained; most of our meteorological instruments and the chemical samplers could operate in rain. With occasional adjournment for special operations, these tests continued for eighteen months to provide a large statistically significant body of data over all four seasons in central Florida. In the Bushnell office, my group translated our field measurements into compact tables of averaged data from the meadow station and from our forest tree station. I describe the high points of final data interpretation in a section below, but first I discuss some events concerning our project that occurred during this period.

COLD WEATHER SPRING

Near Bushnell, cattle farming and vegetable growing were important enterprises. A restaurant in nearby Webster served large, inexpensive, slightly tough steaks. Soldiers and civilians from the CWS project frequently went together for steak dinners. Some of us invited a date to these dinners. Members of our group had become acquainted with other local men and women, and we increased our circle of friends. Our dinner parties became large, numbering about twenty people. In the middle of one large dinner party in early spring of 1945, a man slipped in and talked urgently with a local man at the table, who suddenly stood up and made a sign. About a third of the diners stood up and rushed out. They grew early strawberries, the first to the market sold at high prices, and they risked loss to frost in an effort to be first. The messenger reported that a cold front was on the way, and it would probably produce a heavy frost in our area. The farmers tried to save their crop by covering the plants with straw or fanning them with motor-driven blowers.

RECREATION AND SIGHT SEEING

My girl friend spoke of a rodeo to be held at some nearby place. Four of us went there by jeep one weekend. One scene is shown on the Figure 6.1.A. At another time, I was invited to a special party of the officers. We posed for a picture holding a young boa constrictor, and I was the one to the far right. Figure 6.1.B. On some weekends, two or four of us would go sightseeing in a jeep, and we saw deep crystal springs, lakes, alligator farms, and other nearby Florida sights. Once I went fishing with my sergeant in the Gulf of Mexico. As the boat slowly passed over underwater rock piles, we caught three fine groupers and one big barracuda (Figure 6.1.C). The sergeant took his groupers to the army cook house. I cleaned my two fish and gave them away to two families I knew in town. I visited beaches on both the Atlantic coast and on the Gulf of Mexico.

"LET'S DO IT."

Throughout the project in Bushnell, we followed closely the war in Europe and the Pacific. In Captain Nolen's office complex, Dr. Duncan MacRae daily posted on a bulletin board newspaper stories about the war. Officers and civilians discussed the main battles and speculated about the future. During the allied attack on Normandy in early June 1944, a radio was brought into the office, and it was kept on at a subdued volume for days. The breakout of the allied soldiers from Normandy was celebrated in the office. Later, Dr. MacRae posted war news from Europe on the left-hand side of the bulletin board and war news from the Pacific on the right-hand side. A middle-aged captain in the office, whose name I have forgotten, had been in the chemical warfare service for twenty years before the war started. He took particular interest in the Pacific war. As the United States forces assaulted and recaptured South Pacific islands, he brooded over the large number of

RECREATION AND SIGHTSEEING IN FLORIDA, 1944–1945

FLORIDA RODEO

OFFICERS PARTY with young boa constrictor. I am on the right.

My Sergeant and I with three groupers and one barracuda, Gulf of Mexico.

On Gulf of Mexico beach

Figure 6.1. Recreation in Florida. Pictures given in 1944–1945 to HSJ by unknown photographers.

our casualties. He studied what he could find about the number and nature of the enemy fortifications, made calculations about how much war gas would have been needed to wipe out enemy defenses, and estimated how many fewer losses we would have had. He insisted that we should use war gases in those cases to save American lives. His repetition and insistence finally gave me what I think was an understanding of his position: he craved for war gases to be used for his personal professional fulfillment.

SPECIAL TESTS AND EVENTS

During 1944–1945, sections of NDRC and army groups carried out special tests at the Bushnell site. An NDRC group from a large midwestern university placed its new equipment for measuring mustard along the grid of an upcoming ground-based mustard-gas test. The man in charge was broad and heavy. During the push to get set up before the bomb was burst, the head man accidentally spilled nitric acid on his pants. He quickly and wisely took off his pants and undershorts and doused his legs with water. Time was short, and he ran about directing his crew and setting up equipment dressed only in his shirt and shoes. This event was a subject of jokes for days.

An NDRC group at Harvard had developed a new form of jellied gasoline named napalm. An army unit brought in a fighting tank equipped with a powerful napalm flame thrower, and it maneuvered and practiced in one of our meadows. The napalm weapon was a stream of flaming jellied gasoline that stuck to objects it hit, including people. Foot soldiers with heavy knapsacks attacked trenches and bunkers using napalm, and others used flaming white phosphorus.

Late in the war, a group tested the dispersion of DDT as an aerosol in the meadows and especially in the forest. We made day and night micrometeorological measurements for them.

IDYLLIC TROPICAL ISLANDS OF FLORIDA

In the spring of 1945, Captain Jake Nolen told me that a group from Bushnell was to carry out some special studies on a beach in south Florida. He suggested, and I agreed that I go to carry out basic meteorological measurements. Arthur Pardee was to work on a special project, members of NDRC Division 9 were to make measurements of mustard, and some army officers and enlisted men would be there to handle munitions and other matters.

Before sunup one morning, we were driven to St. Petersburg, which is on the other side of Tampa Bay. We boarded a low, narrow commercial boat and spent the day traveling south. During the early part of the trip, we could see the land of Florida off to our left, but later in the day it was out of sight. After dark, we arrived at our destination and transferred from our boat to a Coast Guard ship, which was tied up at a long pier. Our equipment and personal gear were piled up for the night, and each of us was assigned a place in a three-deck bunk bed.

Some essential army officers had not yet arrived by the next morning, and we were free to explore beautiful and unusual sights. Alone, I walked along the pier and a long sandy beach. Beyond the pier, there was a small island composed of pure white sand with a few palm trees growing on it. Many birds stood on the sand or were flying in the air. I recognized gulls, terns, and pelicans, but there were large birds that I had never seen or heard of before. They were black, except some had a small red patch on the neck. They had exceedingly long narrow wings and deeply parted scissor tails. They flew, glided, and maneuvered beautifully.

I walked off the pier onto the main island (later I was told it was sixteen acres (6.5 hectare) in size). This island was completely surrounded by a brick fort with walls eight feet (2.4 m) thick and fifty feet (15 m) high. Another smaller brick wall formed a moat around the fort. I walked through the entrance into the fort, but inside it was dark. There were large and small boxes of unidentified contents stacked inside.

I learned that we were on the Dry Tortugas Islands. These are half a dozen or so small coral islands that extend far beyond the Florida keys, are seventy miles (112 km) west of Key West, and are ninety miles (150 km) north of Cuba. "Tortugas" is the Spanish word for turtles, and "dry" is the English word that signifies there was no source of fresh water on the islands, except for rain. Ponce de Leon discovered these islands and ate some of its turtles, and pirates heavily used the islands until the early nineteenth century. There had been many shipwrecks on coral reefs in this area. The magnificent black birds were frigate birds, which are common in the tropics. Fort Jefferson was the great brick fort. Construction on it started in the mid-eighteen hundreds, continued for many decades, and was never completed. It was one of a series of large brick forts, which included Fort Sumter in Charleston Harbor, the site of the first battle of the Civil War.

Members of the NDRC group went swimming in the afternoon, and I put on my bathing suit and waded out to waist-deep water. Someone had bought or borrowed a useful item, and it was passed around. It was a one-gallon (3.8 liter) can, painted black on the inside, and had a piece of glass glued over a large hole cut in the bottom. When my turn came, I stuck my eyes and nose into the can, and when I submerged the bottom of the can in water, I could see for great distances through the clear water. Bright colored small fish were swimming in droves. I could see large fish and solid coral formations farther out. The water glowed with white, gray, green, and blue colors.

The next day our work started. By motorboat we were driven to another small island, which was shaped about like a banana. Art Pardee and three soldiers had the job of digging out trenches and building deep defensive bunkers, and they worked hard at this job throughout the day. NDRC men of Division 9 installed samplers to measure the amount of mustard that would be deposited. I set up the micro-meteorological station, which was a small version of our stations in Bushnell, and I and made measurements throughout the day. Between times, I walked up and down the island, which was all pure white sand, had two kinds of palm trees, and to our surprise had lots of cactus. Many birds were in the trees and on the ground. The water showed blue and green patches. I think the army planned to bring out goats the next day, but I had not seen any.

In the afternoon, another boat arrived with the commanding officer for this operation. He was a high officer in the chemical warfare corps, and he came directly from Edgewood Arsenal, the Chemical Warfare Service (CWS) headquarters. The officer sat in his boat fishing. From time to time his sergeant, as instructed, drove the boat to other points along the shore. He never got out of his boat to inspect the soldiers' work and never spoke to anyone on the island. Finally, the officer and his sergeant went back to the fort.

We returned at the end of a full day's work. At dinner that night, an officer told me that the high officer had examined the site and found no need for meteorological measurements. The next day three soldiers and I took a small fast boat to Key West, and we went back to Bushnell in a six passenger weapon carrier.

Shortly after Arthur Pardee came back to Bushnell from the Dry Tortugas Islands, he left and joined a war project in the Radiation Laboratory at Berkeley. Captain Nolen was furious. He said Pardee had abandoned real war research at Bushnell and fled to ivory tower research in the University at Berkeley. Only later could it be said that Pardee had left Florida to work at Berkeley on the nuclear bomb program.

Interpretation of Observed Meteorological and Weapon-test Data

AREAS COVERED BY VARIOUS AMOUNTS OF MUSTARD GAS

Captain Nolen and his advisers planned a series of ground-based, single-bomb tests to obtain some empirical information about how mustard gas clouds behaved in terms of meteorological and micro-meteorological conditions. Duncan MacCrae and a group of army officers took chemical analytical data from the NDRC group and placed the numerical value of each observed dosage on a large map, which showed where chemical samplers were placed. They found that, in general, the logarithm of the distances from the bomb source to the samplers was linear with the logarithm of dosage along radii from the source, and they interpolated and entered on the map the dosage values between source and each sampling station. For example, suppose the dosage at the bomb site was 114 (in some units of measure) and the dosage at a station was 37; by their formula concerning logarithms, they entered on the map along a line between source and station where the values would be 100, 75, 50, and 25. Suppose at another station at a different distance from the source and at a different angle, the measured dosage was 13. MacCrae and his group would locate on the map where the values would be 100, 75, 50, 25, 10, and 5. After this procedure was carried out for each chemical sampler, one of the men drew an ellipse connecting all values of 100 and another ellipse connecting all values of 75 and so on as far as samplers went. Figure 6.2 gives an idealized example of what the contour maps were like. By use of a simple tool, a planimeter, they found the area inside each contour loop. They tabulated their results as the area A within which the dosage was ***D***[#] or more. Every week, Captain Nolen submitted these results to Edgewood Arsenal, the national headquarters of the Chemical Warfare Service, and these reports included micro-meteorological data and verbal descriptions of the cloudiness and the weather.

[#]Dosage, *D*, is defined as the product of gas concentration in air multiplied by the time an animal or person is exposed to the concentration. The chemical samplers drew in outside air at a steady known rate for a known time and extracted the mustard gas from the air. The total amount of mustard gas in the sample gave dosage at the location of the sampler. Also, there were other sampling devices in use.

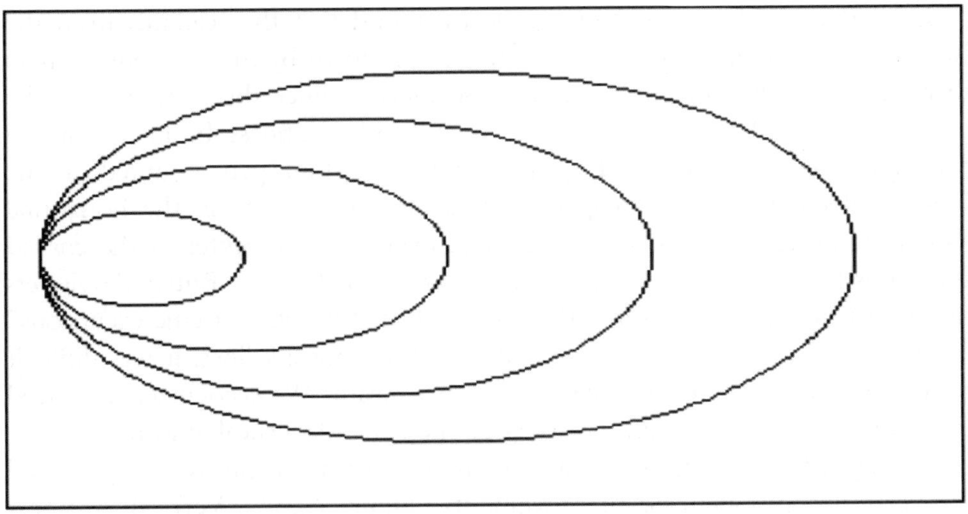

Figure 6.2. Idealized contour map showing areas (***A***) covered by gas dosages (***D***) at four decreasing values of ***D***. Dosage is the product of gas concentration ***C*** multiplied by the time ***t*** summed over all the time of a test. One property of a gas mask is the dosage it will retain.

Prepared by author for this book.

OUR RECORDING GUSTINESS METER

Early in 1945, Captain Nolen allowed my group and me to examine the dosage and area data. After studying the contours of area and dosage of the gas, we were able to write in our report[§]:

"It was observed that for each operation the area ***A*** covered by at least the dosage ***D*** could be represented by the equation

$$\text{Area } A = K/D^n$$

over all the target grid except for a small area close to the source. The terms ***n*** and ***K*** vary from operation to operation and are functions of meteorological conditions, terrain, and other conditions."

[§]"A study of turbulent diffusion of gas clouds over several terrains," Report OSRD No. 6185, 84 pages, by Harold Johnston, Robert Merrill, and Robert Mills. 1945. Part I. An Empirical Approach to the Effect of Turbulent Diffusion on Gas Clouds over Several Terrains, pages 1–23. Part II. A Critical Examination of the British Statistical Diffusion Theories, pages 24–46.

In working with the contour maps, we found that the readings from the British bivane correlated better with the spread shown by the gas contour map than our other micro-meteorological observations. Since the early days in the cow pasture at Pasadena, we had used a British gustiness bivane. It was a windvane, and it also swiveled up and down. A red-ink pen mounted on the vane scribbled on a curved paper as wind moved the vane in the horizontal and vertical directions. After a two-minute exposure, the center of the curved paper was coated by red ink tracks. As recommended by the British developers of this instrument, we measured the widths of the "average extreme excursions" in the vertical direction. From the coil of overlapping ink lines, it was difficult to assign a width or to say what the record meant. We recognized the need for a quantitative, physically meaningful measure of vertical gustiness.

I designed an instrument that measured vertical gustiness and provided the vertical component of air movements (Figure 6.3.A). Vertical gustiness (***Gz***) is a ratio of turbulent vertical velocity divided by wind velocity.

$$\boldsymbol{Gz} = \frac{\text{vertical wind}}{\text{wind}}$$

One vane of the gustiness meter aligned the system with the wind; the other vane freely swiveled up and down on high-quality jewel mounts. The tangent of the angle of the vane from the horizontal plane was the instantaneous gustiness as defined above. The vertically moving vane was made of balsa wood and tissue paper, and its streamlined shape gave laminar air flow around it. Electrical contacts were maintained between the movable vertical vane and a stationary milliammeter, which read the value zero for an angle of 60 degrees downward, 0.5 when the vane was parallel with the earth surface, and 1.0 when the vane was displaced upward by 60 degrees. By throwing a switch, we could change the range between 0 and 1.0 on the meter to be minus to plus 120 degrees. With another instrument we measured the average wind speed over the same five minutes, and we operated both instruments at a height of two meters above the ground. On the glass face of the milliammeter, we pasted two scales that gave the gustiness directly for each of our two sensitivities. One of us read the instantaneous gustiness every five seconds for five minutes and another wrote down the values. We averaged the positive deflection to give the average positive vertical gustiness. In some cases we recorded the gustiness readings on the Esterline Angus milliammeter and used a planimeter to measure the area for the positive excursions of the windvane. We multiplied the average positive vertical gustiness by the average wind speed to give an estimate of the vertical turbulent velocity.

$$\text{vertical wind velocity} \approx \text{average gustiness} \times \text{average wind velocity}$$

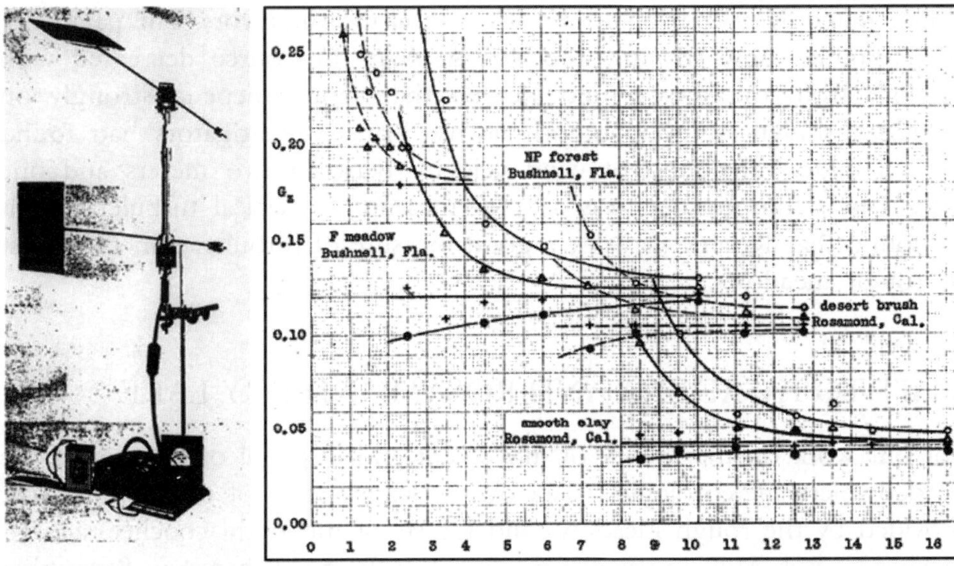

A. Vertical gustiness meter. **B.** Wind speed at 2 meters above ground/miles per hour.

Figure 6.3. Vertical gustiness plotted against wind speed observed at two meters above the ground for four locations and at various degrees of thermal stability and instability.
[Note] Symbols on Figure. T(2m) – T(0.3): ○, –2.0; Δ, –1.0; +, 0.0; •, 0.5.

Picture taken by unknown army photographer; figure drawn by unknown army officer, 1945.

The machine shop at Caltech built several copies of our new meter. We used it in the Bushnell forest and meadows, and obtained some interesting results.

* * *

We produced what I regard as a high-quality scientific report,[§] which was classified as CONFIDENTIAL, stored away, and forgotten. In 1997 I obtained a copy of this declassified report through the Freedom of Information Act, and most of this section is abstracted or quoted from that report. The object of Part I was to relate empirically the readings made by our new gustiness instrument, which gave us the vertical turbulent wind velocity, to the spread of gas clouds over a wide range of terrain. The object of Part II was to apply the general British statistical diffusion theory for a continuous point source of vapor. In our case the "point" source was an evaporating pool of liquid mustard gas formed by a bomb exploded on the ground. We sought to relate the parameters for spread of the gas cloud, ***n*** and ***K***, to our micro-meteorological measurements. The parameter ***n*** determines how strongly the dosage near the source decreased with distance from the source, and we felt that should depend strongly on vertical turbulence. Over smooth lawns, British investigators had found the "R-value," the ratio of wind velocity between two meters and one meter above the ground, was a good measure of vertical turbulence, but we found that R-value was not a good measure of turbulence in the forest or rough meadows.

VISIT TO PASADENA AND ROSAMOND DRY LAKE

For our work on the theory of movement and dispersal of gas clouds, we needed micro-meteorological data over a wider range of terrain than that provided by the rough meadows and forests of the Withlacoochee stations in Florida. Bob Mills returned from San Jose Island, Panama, in September 1944 (see Chapter 5) and worked on Pauling and Corey's project at Caltech; that project concerned the chemistry of rocket propellants. In the late spring of 1945, I arranged to take two weeks off from Bushnell, and Dr. Corey let Bob take two weeks off from his project. I went to Pasadena, Bob and I borrowed a Caltech panel truck, and on two trips we camped out at Rosamond Dry Lake for several days, while taking many measurements including our new electrical bivane. We saw no one at the dry lake during either trip.

This two-week working vacation at Rosamond Dry Lake was a pleasant experience. At sunrise, air in the desert was chilly, by mid-morning it was quite warm, by noon it was hot, and we dressed accordingly. We used three locations for making measurements, one with smooth flat vegetation-free surface, one with a surface of cracked dry clay with upturned edges about one inch (2.5 cm) high, and the other in desert brush. Bob was setting up our apparatus in the chilly early morning in Figure 6.4.A, and the wide smooth surface of dry lake bed was in the background. We traveled back and forth between the three stations by bicycle, Figure 6.4.B, and a sample of Mojave desert brush is in this background. Between the two trips to the desert, we built a sail for the bicycle; in Figure 6.4.C I was riding the sailbike over the cracked curly clay tiles of the desert. It was windy day and night, especially in the afternoon, and we noted that our eight-foot (2.4 m) ladder blew over when the wind reached twenty-five miles (40 km) per hour. We slept outside on army cots. The Caltech panel truck was tall enough for us to stand up straight, and we used the truck for our kitchen,

A. Bob Mills setting up micro-meteorological equipment, early morning, long shadows.

B. Bob commuting between stations on bike Dry Lake brush in background.

C. Hal on sailbike on cracked, curled, clay surface.

Figure 6.4. Rosamond Dry Lake, 1945.

Photographs by Hal Johnston and Bob Mills.

including a camp stove and icebox. We replenished the ice at a filling station in Lancaster. Mostly we cooked our meals, but occasionally we ate in Rosamond, sharing the food-counter with cockroaches.

PROFESSOR ROSCOE DICKINSON IN 1945

Before I arrived at Pasadena, Professor Dickinson had undergone surgery for colon cancer. He was out of the hospital, at his home with Madeline, and I was told that his prognosis was favorable. Bob Mills and I visited him. He was in a wheel chair, but he was cheerful and joking as usual. I told him about our work in Florida and our pine-tree meteorological tower. Roscoe and Bob chatted like the old chums that they were.

In July 1945, Professor Roscoe Dickinson died.

VERTICAL GUSTINESS

Figure 6.3.B contains a large amount of information. It shows how vertical gustiness varies with wind speed, vertical temperature gradient, and terrain. The four sites were the smooth flat Rosamond Dry Lake, flat brushy desert, an irregular bushy Florida meadow, and the dense Florida forest. At zero wind speed under lapse† conditions, large sized convection cells formed, and there was strong vertical mixing and large vertical gustiness. As wind speed increased, the convection cells were distorted and displaced, and turbulence was mechanically induced by friction of the wind against surfaces. This mechanically induced turbulence tended to break up the large convective action, and the hot air, instead of passing as a steady stream up the middle of convective cells, had to fight its way upward as isolated streamers and bubbles. As the wind speed increased to large values, the mechanically induced turbulence became greater than the turbulence produced by the hot air bubbles, and gustiness asymptotically approached the thermally neutral value. Over the smooth lake-bed surface, it took a higher wind speed to produce enough mechanically induced turbulence to break up the convective cells than it did over rough surfaces (Figure 6.3.B).

At low wind speed at night, the surface was cooled by emission of heat radiation, air was cooled upon contact with the surface, and cold dense air pooled on the surface as if a separate fluid. We had an inversion. Where the ground sloped, the cold surface air flowed downhill, a katabatic wind, and mechanical turbulence was induced by this wind. When there was a prevailing wind, its mechanically induced turbulence was reduced by the

†I use "lapse" here as abbreviation for unstable temperature lapse rate, "inversion" for temperature inversion, and "thermally neutral" for zero-temperature gradient.

temperature inversion. At high wind speeds, the mechanical turbulence was only slightly reduced by the inversion, and the measured gustiness approached the thermally neutral value.

The lines through the plus marks (+) in Figure 6.3.B were those of zero temperature gradient, and they gave the purely mechanical gustiness. For each location, this mechanical gustiness was the same at all wind speeds, but this quantity was different for each terrain. Where vertical gustiness is constant with wind speed, the vertical turbulent wind is directly proportional to wind speed. The mechanical gustiness was largest in the forest, was much less in the Florida rough bushy meadow, was somewhat less in the desert brush at Rosamond Dry Lake, and was much less for the smooth clay lake bed at Rosamond.

OUR INTERPRETATION OF THE FIELD SAMPLING DATA

Various investigators had conducted field experiments with chemical warfare munitions over a wide variety of terrain, latitude, and climate. In 1945 there was no adequate systematization of these data, an inevitable outcome of non-standardized sampling and reporting.

During 1945, my group took the contour maps of mustard gas dosages that were based on single bomb tests statically fired, which we regarded as a point source in terms of the British statistical diffusion theories of G.I. Taylor and O.G. Sutton.‡ We paired these contour maps with our micro-meteorological data. Through the library at Edgewood Arsenal, we obtained gas measurements in the field and micro-meteorological observations from the Canadian station at Suffield; these studies had been over smooth snowfields and over prairie covered by short grass. We also obtained limited amounts of such data from San Jose Island of Panama, Australia, and Britain.

Under the range of wind speeds and temperature gradients available in Florida, we compared, side by side, our gustiness meter with the British bivane in the Bushnell meadows and forest. Also we made these comparisons at Rosamond Dry Lake. At all of these sites and with large scatter of points, we found that standard vertical width of the bivane record multiplied by the numerical factor of 0.23 was equal to our measured vertical gustiness:

British bivane measure × 0.23 = Our measured vertical gustiness

Table 6.1. Zero Temperature Gradient Values of Mean Vertical Gustiness ***Gz*** for a Wide Range of Terrain and Exponential Factor ***n*** for Dosage Areas. Table IV of reference.[§]

Brief description and Location	***Gz***(obs)	***n***(obs)
Flat smooth snow field, Suffield, Alberta, Canada	(0.044)	1.05
Flat smooth hard clay, Rosamond Dry Lake, Calif.	0.044	
Smooth Gulf of Mexico, Dry Tortugas Island, Florida	0.048	
Cracked clay, upturned edges, one inch (2.5 cm), Rosamond Dry Lake	0.060	
Grass 2 inch (5 cm) high, prairie, Suffield, Alberta, Canada	(0.066)	0.88
Grass, 6 inch (15 cm), some foot-high weeds, T-meadow, Bushnell, Fla.	0.083	0.85
Desert brush, 20% cover, 20 inch (50 cm) high, Rosamond Dry Lake	0.105	
Rough meadow, 6 inch (15 cm) grass, some bushes. Bushnell, Fla.	0.12	0.77
Dense semi-tropical forest, 60 foot high (18 m) canopy. Bushnell, Fla.	0.18	0.68
Tropical jungle, San Jose Island, Panama	(0.20)	0.65

Our gustiness meter measured the angle of the vertical displacement from the horizontal. The British bivane measured the full width of the vertical ink tracings, which is twice the upward directed component that we used, and the bivane was smaller and had a much shorter length from vane to swivel point than our instrument. These two features explained why the readings of the two gustiness meters differed by a constant value. For the rest of this chapter, I write within parentheses the British bivane measurements multiplied by the factor 0.23 for measurements not made by our group.

We prepared Table 6.1, a compilation of observed thermally neutral gustiness and of the exponent ***n*** derived from field studies of war gases for sites that varied from smooth flat snowfields to a tropical jungle. Each successive entry in the table:

(i) corresponded to terrain recognizably rougher than the entry above,
(ii) had observed gustiness value larger than that above, and
(iii) gave a smaller value of the cloud spreading parameter ***n***.

A conclusion from this table is that thermally neutral vertical gustiness is a unique measure of the roughness of a site. This table suggested that we plot the parameter n for all available munitions data against observed gustiness, including unstable lapse and temperature inversion conditions, Figure 6.5. There was surprisingly good correlation between measured vertical gustiness and the parameter for area sampling, but the data available for this plot did not include any large temperature inversion.

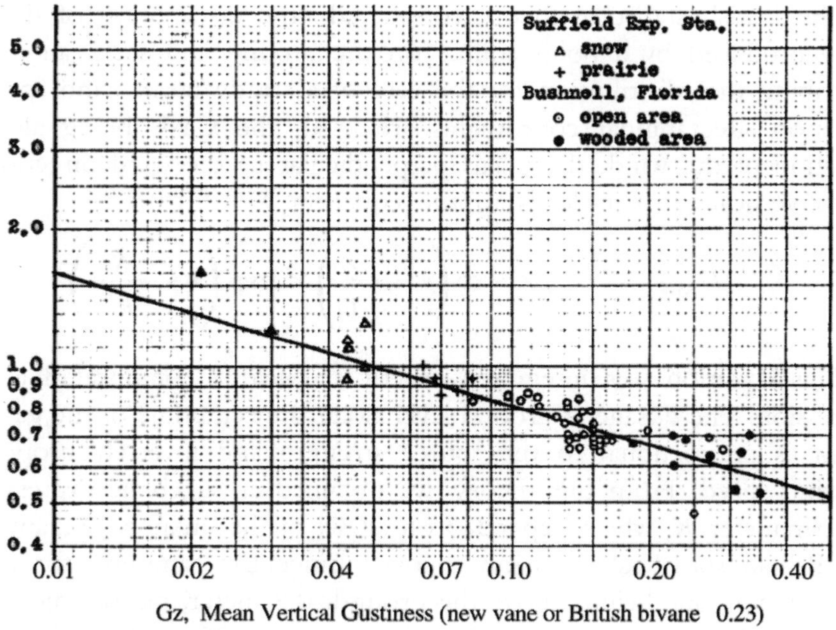

Figure 6.5. The Relation of Mean Vertical Gustiness to the Parameter n Descriptive of Area Sampling, for a variety of Terrain and Meteorological Conditions. Figure IV of reference.[§]

Gustiness is the ratio of turbulent vertical velocity and wind velocity. Recall the relation: Area $A = K/D^n$. For two areas A_1 and A_2 with corresponding dosages D_1 and D_2:

$$\frac{A_2}{A_1} = \left(\frac{D_1}{D_2}\right)^n$$

the parameter n determined the ratio of the area covered by a dosage of gas relative to the area covered by another dosage. Gustiness gave no information about the magnitude of dosage D anywhere. The term K determines the magnitude of the dosage within the source region, which depended on source strength, wind speed, and other features.

During 1945, my group sought to interpret the gas-cloud data with the British statistical diffusion theories of G.I. Taylor and O.G. Sutton.[‡]

‡G.I. Taylor, "Diffusion by Continuous Movements," *Proc. Lond. Math. Soc.* **20,** 196, 1922. O.G. Sutton, "A Theory of Eddy Diffusion in the Atmosphere," Proceedings of the Royal Society of London, **A135**, page 143, 1931; and "Diffusive Properties of the Lower Atmosphere," MRP 59, 1942.

The British statistical theory considered "the distribution of dosage in space produced by a point source on the ground and seeks an equation describing that distribution for the case of a steady mean wind speed and direction. Rectangular coordinates are used, the x-axis being along the direction of the wind, the y-axis perpendicular to the wind direction, and z-axis vertical ... This leads to the equation,"[§]

$$D_{x,y,z} = D_{x,0,0} \times \exp\left(y - \frac{r}{F_1}\right) \times \exp\left(z - \frac{S}{F_2}\right)$$

This equation was integrated exactly to yield a complicated algebraic expression. The dosage at ground level along the centerline of the gas cloud, $D_{x,0,0}$, is proportional to the source strength and inversely proportional to wind speed. From this theory, we obtained an algebraic expression for the parameter K, which depended on all six of the unknown parameters of the British statistical diffusion theory.[‡] Since the sampling for gas was done at only one height above the ground, 0.3 meter, measurements made at Bushnell were not enough for us to calculate K.

When we showed these results to the recently promoted Major Nolen, he became quite excited. He pointed out that if we had a satisfactory way to estimate the parameter K, we could extrapolate the munitions tables obtained at Bushnell to all the sites of Table 6.1. He urged us continue to think about the problem. We thought about using Kenneth Pitzer's pancaking theory (Chapter 2) to solve for an initial area and dosage, and then switch to the statistical diffusion theory and Table 6.1 to complete the calculation.

Last Phase of Our Work

In the summer of 1945, my housemate and colleague, Second Lieutenant Alan Englander, married a young woman who was a native of Bushnell. His family in New York came down for the wedding, and many of her large family attended. The church was full and hot for the wedding.

WAR SECRETS AND GOSSIP

Many civilians and officers visited our project in Bushnell. Visitors included professors and business officers from universities that had NDRC projects, officers in NDRC, army liason officers, representatives from other divisions of NDRC, and officials of various sorts from Washington. A few of these visitors wanted to see our meteorological systems, and I frequently drove such visitors out to see our instrumented tree and other equipment. A small number wanted to see our micro- meteorological results. I talked with the visitors and listened to them. There were so many people of this kind that I do not remember who told me some of their stories, although I do remember about when and what was said. One visitor said in the spring of 1945 that our army had developed a powerful new explosive, and in its test it had blown up a mountain in Tennessee. Later another visitor, seeming to believe that I knew what he was talking about, said that we had switched from research to manufacturing. From context, I concluded he was talking about the new explosive that blew up a mountain. I had SECRET security clearance, but I had no "need to know;" these communications were illegal. In the Caltech Athenaeum in 1942, I learned something about nuclear fission from conversation among physics graduate students. I did not figure everything out concerning nuclear bombs, but I sensed that something really big was developing in this area.

FINISHING UP

In the summer of 1945, I took a train to Washington, DC and to Edgewood Arsenal to search their library for material in order to estimate the initial area and gas concentration after a bomb exploded . While I was there, U.S. aircraft dropped a nuclear bomb on Japan. As my return train approached Florida, a hurricane swept over Bushnell and blew down trees over a wide area, including one large oak tree in front of the house we were renting (Figure 6.6.). After the second nuclear bomb was dropped, the war ended. After a couple of weeks, I submitted our large report[§] to Major Nolen, took a week's vacation to visit my family in Georgia, and returned by airplane to California.

Throughout my stay in Florida, I was paid a salary of $125 per month, and also I received $6 per day, with which to pay for room and food. In the twenty-two months I spent in Bushnell, I saved three thousand dollars, which was a large sum of money at that time and in my group of acquaintances. I later used it as down payment on our first home.

I returned to graduate school at Caltech in the fall of 1945. Professor Don Yost became my new research director.

My last act of chemical warfare occurred about a month after classes started in the fall of 1945. Mr. Matt Thompson was in charge of Caltech grounds, and he was given the task of disposing of toxic chemicals from Yost's and Niemann's laboratories, many of which included small amounts of war gases. Niemann suggested that Thompson should contact me. I

Our rented house, Bushnell, Fla., 1944–1945.

Wildwood, Fla., Hurricane, August 1945.

Oak tree blown down in front of our rented house by the hurricane, August 1945.

Figure 6.6. Photographs by Hal Johnston with an old used camera, bought in August 1945.

still roomed in Mr. and Mrs. Thompson's house. We talked about the problem, and I proposed a solution. Caltech owned some land well out in the desert. Thompson and his crew went to the desert and dug a hole six feet (1.8 m) long, three feet wide, and six feet deep, and they rolled a large stone into the pit. Matt and I rode out in his pickup truck, and in the back of the truck there were cardboard boxes containing a number of glass vials and a few bottles, which I had carefully packed. From the quantities involved, I could see there was no real danger. I put on a gas mask, removed the boxes from the truck, and put them down beside the pit. Matt promptly started up the truck and drove it about a hundred yards (90 m) away. I picked up the glass containers and tossed them hard against the big rock in the bottom of the pit. As I recall the glass broke every time. I could read the labels on the vials and bottles, and I orchestrated the pitches so as to maximize the smoky fumes that came out to the top of the pit. Occasionally I glanced at Matt, amused to see that he was terrified. After I finished, I walked up to his truck. He said he would get his crew to fill the hole with dirt and drove me back to town.

DIAGNOSIS

In this book, I quote doctors who said I could expect to live only until I was thirty or maybe forty years of age. Here I am writing this book at age eighty. What didn't happen? My heart problem has been a domineering feature of my life, but it was correctly diagnosed for the first time when I was fifty-five years old: I have congenital mitral valve prolapse (MVP) with a moderately severe "leak" (doctors call it "regurgitation") from early childhood. It is a mechanical defect that worsens slowly, over many decades. It was severe enough to cause shortness of breath when I ran at age eight, but doctors did not recognize it. Sharp chest pain, which I had for many years, is a common effect of MVP.* At age thirteen I had a full case of rheumatic fever that upset the rhythm of the heart and did generalized irritation and damage. Mitral valve prolapse was unknown at that time. The doctors thought the arrhythmia, fast pulse, heart leak, and heart

*Harislos Boudoulas and Charles F. Wooley, *Mitral Valve Prolapse and the Mitral Valve Prolapse Syndrome*, Futura Publishing Company, Inc., Mount Kisco, New York, 675 pages, 1988.

enlargement were all due to rheumatic fever, in which case my prognosis was poor. The doctors had often seen such damage from rheumatic fever develop into mitral valve stenosis (narrowing of the opening) and fatal congestive heart failure within a few years. After about age eighteen, I exercised and worked as hard as I could without undue discomfort. It was not a good idea for me to run, I never engaged in competitive sports beyond ping-pong, and I could never swim. By age thirty I could hike four miles (6.4 km) and slowly walk up a thousand feet (300 m) per day in the Sierra Nevada mountains. Mary Ella Stay said she would be happy to take her chance with me. When I was twenty-eight, Mary Ella and I were married, and we have four children. The doctors were wrong there.

Thoughts about War Gases in World War II Based on Specific Cases in this Book

"DRAFT DODGERS"

In July 1944, while Bob Mills was working with the army in the jungles of Panama, he received a notice from his draft board in Ohio to report for induction into the army. A high army officer at San Jose Island telephoned an appeal to the draft board, and they granted him a deferral until he returned to the United States. Pauling at Caltech made a strong appeal for Bob, and high NDRC officers in Washington also took up the case. Eventually the draft board withdrew the induction notice and gave him a conditional extension.

The official position of NDRC on the draft was expressed by W.A. Noyes, page 8:

> As the demands of the Armed Services increased, the danger of losing scientists through Selective Service also increased. And yet the success of NDRC and of the war effort in general depended on retaining scientific men in the laboratory. Many of these men were young and vulnerable to the draft. A large amount of time was necessarily spent in ensuring deferments of scientists, and although NDRC lost only a few men, these duties were very time-consuming and annoying.

I believe that most of the young people that I knew in NDRC wanted not to be drafted to serve in the armed forces. I believe that the army and navy wanted the National Defense Research Committee (NDRC) to employ technically skilled civilians to "conduct research for the creation and improvement of instrumentalities, methods, and materials of warfare." In large numbers, young scientists dropped their interesting studies and research to join NDRC projects, satisfying their personal preference for not joining the armed forces and fulfilling the need of the armed forces for high-quality, sometimes creative, research and development.

From the list of names in Noyes' book, I estimate there were about seven hundred scientists working on NDRC chemical projects. At the end of Chapter 3, I reproduce Noyes' list of twelve chemists who died in the action of doing their NDRC jobs, which is a fatality rate of 1.7 percent. Two of the twelve died from war gases in laboratory work, two from work with high explosives in the laboratory, one while working on underwater explosives, and one as his blimp crashed while he was working on underwater flash illuminations. I was unable to find what work the other six were doing. This death rate is small compared to that of Air Force bomber crews and to soldiers and sailors invading Normandy, in the Battle of the Bulge, retaking Pacific islands, etc. For comparison with military service as a whole, the Encarta Encyclopedia states that the United States mobilized sixteen million people for military service and sustained 292,131 battle deaths, which is a fatality rate in battle of 1.8 percent, only slightly more than that for NDRC chemists. Upon reading Noyes' full book, I conclude that the seven-hundred NDRC chemists contributed much more to the war effort than they would have if they had been in the armed forces, and the young scientists involved were aware of this fact. I think the epithet "draft dodger" is not applicable to NDRC personnel.

DEFENSE OR OFFENSE

Professor Roscoe Dickinson was a brilliant scientist, an artist, and a generous liberal person. He felt good about working with terrible poisons to provide better gas masks. When NDRC Division 10 shifted its emphasis on chemical warfare from defense to offense, he adjusted to the change and supported the new emphasis, accepting that a recognizable capability to go on the offense was a necessary part of defense.

WHO MIGHT HAVE WANTED TO USE WAR GASES?

As noted in Chapter 6, a long-term professional chemical warfare officer "craved for war gases to be used, for his personal professional fulfillment." An expert's pride in his or her own specialty can be a strong motivating factor and possibly lead to biased advice.

Although Allied toxicologists and organic chemists tested over two thousand compounds to see if any would make a promising war gas, they did not discover sarin and other nerve gases, but the Germans discovered and stockpiled large amounts of nerve gases. Sole possession of this secret weapon might have tempted some in Germany to want to use nerve gas, for example, to help repel the Allied attack on Europe in 1944.

In 1942 (Chapter 1) Dickinson's group tested di-methyl fluorophosphate and di-isopropyl fluorophosphate with respect to their retention by charcoals. The nerve gas, sarin, is close to being intermediate in structure between the two we worked with.

<table>
<tr><td>O
||
CH3 – O – P – O – CH3
|
F</td><td>O   CH3
|| |
CH3 – P – O – CH
| |
F   CH3</td><td>CH3   O   CH3
| || |
CH – O – P – O – CH
| | |
CH3   F   CH3</td></tr>
<tr><td>di-methyl
fluorophosphate</td><td>sarin</td><td>di-isopropyl
fluorophosphate</td></tr>
</table>

NDRC Division 9 projects at the University of Illinois and the University of Chicago synthesized a large number of fluorophosphates (Noyes, page 168). Since sarin is so close in molecular structure to the fluorophosphates that we had (chart above), it seems to me that NDRC probably synthesized sarin and tested it on animals. However, fluorophosphates showed greatly different toxicity from one specie of animal to another, and the uncertainty about toxicity to humans could have been the reason that the Allies did not identify any of the fluorophosphates as possible war gases. Nazi Germans are known to have used human subjects to test the lethality of war gases. [O. Bichenbach, Deposition before French Military Tribunal, May 6, 1947 (translation by Office of the Chief of Council for War Crimes, US Army) Tribunal Militaire de la Gieme Region, Document No. 1852.]

Early in 1944, U.S. and British forces knew that the quick killer cyanogen chloride readily penetrated German and Japanese gas masks, and Allied gas masks successfully stopped this gas. This advantage might have tempted Allied forces to use cyanogen chloride, for example, against Japanese soldiers occupying islands in the southwest Pacific.

After the war, General Bullene wrote a letter to Professor Blacet and to all of the NDRC group who went to Panama: "As you know the reports and tests were finished in time so that they could have been used had our specialty been called for when the Japanese homeland was invaded." What a euphemism, "our specialty !" At that time and place, Allied forces knew that we were ready to use war gases, and the other side probably knew it also.

Soldiers clad in full protective clothing in 1944 survived and operated in an area heavily contaminated with mustard gas for at least twenty-four hours. Under summer or tropical conditions, the soldiers were intensely uncomfortable and of reduced capabilities. Some were injured by the mustard (see Chapter 4). If one side in the war had initiated use of poison gases, the other would soon have retaliated, and both sides would face the hazards of gases and the burden of full protective clothing. Throughout the war, use of gas offered no overwhelming advantage to either side, and it would have greatly increased the supplies required by the armed forces on both sides.

On the steps of Gates Chemical Laboratory at Caltech in 1942, the Colonel told me that, compared to a massive attack by aircraft delivering phosgene bombs, there would be "a bigger pay-off if explosives were used."

My specialty, micro-meteorology, provided an inhibition against use of war gases in World War II. The lethality of a given mass of war gas varies strongly on ground-level wind speed and vertical-temperature profile; at low winds a factor of a hundred difference might be expected between a moderate temperature inversion and a daytime unstable vertical lapse rate. These considerations cause large uncertainty in the efficacy of war gases.

In Conclusion

Early in 1942, Professor Roscoe Dickinson said to me,

> **"We are guarding a bridge that may never be attacked; we hope it will not be. If it is not attacked, our work has succeeded."**

The bridge was not attacked, that is, war gases were not used. Thus I assert that Professor Dickinson's work succeeded.

CHAPTER 7

OTHER DIVISIONS OF NDRC

Breadth of Activities of National Defense Research Committee (NDRC)

This book gives a deep, narrow view of some activities of NDRC, Division 10, during World War II. It concerns a small number of projects and small number of research workers, and it gives details for only two universities that participated in this war work, Caltech and the University of California at Berkeley. In this chapter, I give a brief examination of the breadth of the National Defense Research Committee (NDRC).

"On 27 June 1940, leading scientific statesmen and knowledgeable civilian political figures established the National Defense Research Committee (NDRC)." (Quoted from Preface of this Book.) In June 1941, the leadership of NDRC established a broader organization, the Office of Scientific Research and Development (OSRD), which was in charge of NDRC and other fields of civilian war research. Vannevar Bush was director of OSRD, James Conant was chairman of NDRC, and W.A. Noyes, Jr. was head of Division 10. A set of monographs was published in the late 1940s, *Science in World War II*, including four volumes devoted to NDRC:

(a) *Rockets, Guns, and Targets*, Divisions 1, 2, and 3 of NDRC;
(b) *New Weapons for Air Warfare*, Divisions 4, 5, and 7 of NDRC;
(c) *Chemistry*, Divisions 8, 9, 10, 11, 19, [20] of NDRC, W.A. Noyes, Jr., Editor;
(d) *Applied Physics: Electronics, Optics and Targets*, Divisions 13, 14, 15, 16, 17, 18.
(e) *Advances in Military Medicine*, Committee on Medical Research.
(f) *Combat Scientists*, Office of Field Service.
(g) *Organizing Scientific Research for War*, Administrative framework of OSRD.

The book (c), which is devoted to chemists' contributions to NDRC, includes thirty-three chapters grouped into six parts:

(i)	Division 8,	Explosives;
(ii)	Division 9,	Chemistry;
(iii)	Division 10,	Adsorbents and Aerosols;
(iv)	Division 11,	Chemical Engineering;
(v)	Division 19,	Miscellaneous Weapons;
(vi)		Tropical Deterioration Committee.

These titles fail to show the breadth of the Divisions activities.

SOME MAJOR ACHIEVEMENTS

Throughout this book I have liberally quoted the monograph (c) on chemistry, which W.A. Noyes, Jr. edited. In the Foreword of Noyes' book, pages xii and xiii, James B. Conant, Chairman of the National Defense Research Committee and President of Harvard University, and Roger Adams, member of National Defense Research Committee, stated their opinion as to the major achievements of the chemical divisions of NDRC:

> The first efforts of the Committee after its establishment in June 1940 were directed toward discovering what were the most pressing problems. Yet even the Army and Navy were reluctant at that time to disclose to a civilian organization all the results of the confidential investigations in their laboratories; rather, they tended to confine their suggestions, as far as chemistry was concerned, to a few fundamental problems of somewhat academic character. These and other proposals by men who had had contact with Army and Navy research were assigned by NDRC almost exclusively to university laboratories where facilities and competent staff members were available.
>
> After December 1941, however, the picture changed completely. Industry co-operated to the fullest extent and accepted research problems whenever properly equipped laboratories and trained personnel were available. The Army and the Navy revealed to the civilian scientists their important requirements in the chemical field. The NDRC program expanded rapidly.
>
> Certain Army and Navy officers were skeptical about the ability of academic scientists to contribute in an effective way. In the field of explosives, particularly, the tradition had been that no individuals lacking special experience and training in this field could be successful. . . Actually, a new method for producing RDX was developed, based on the work of university men who had had no previous experience in explosives.
>
> Among other important accomplishments of the chemists mobilized through NDRC we must mention the discovery of hydraulic fluids, with

very flat viscosity curves over a wide range of temperatures, which greatly simplified the problems of the Air Forces; the determination of a practical method for thickening gasoline, which resulted in revolutionary improvements in flame throwers and made possible the design of new oil-incendiary bombs; the perfection of absorbents in gas masks, both activated carbon and filters, which supplied protection for United States troops far superior to that of the enemy forces. All of these developments originated in university laboratories. Oil smokes for screening large areas during the war stemmed from the application of the basic principles of physical chemistry.

RELATIONSHIPS OF NDRC PERSONNEL TO THE ARMED FORCES [Noyes, pages 141–152].

The relationship between NDRC on the one hand and the Army and Navy on the other constitutes one of the most important phases on the history of the scientific effort of the United States from 1940 to 1946. Many mistakes were made and at times there was much bitterness . . . Many NDRC representatives were irked by the long lapse of time between the inception of an idea and use of a developed item in Theaters of Operations. [Noyes, page 141].

Let us outline a typical procedure for carrying research from the brain to the field. The steps may be tabulated as follows:

Covering more than one full page, Noyes gave twelve detailed steps. [Noyes, pages 142–143].

The . . . Technical Division was constantly irritated by attempts of NDRC to force adoptions at higher echelons and its too great willingness to forget military channels and military courtesy. [Noyes, page 147].

. . . it is only fair to say that the personnel dealing with this subject from the Army side were not properly qualified. [Noyes, page 152].

As might be expected, relationships were good or bad depending on the personalities of the individuals involved. [Noyes, page 152].

Also, as might be expected, some NDRC activities experienced breakdowns and failures.

* * *

Members of NDRC Division 10 carried out "the perfection of absorbents in gas masks, both activated carbon and filters, which supplied protection for United States troops far superior to that of the enemy forces" and also "oil smokes for screening large areas." Division 8 developed "a new method for producing RDX" and made other major improvements on explosives. Division 11 discovered "hydraulic fluids, with very flat viscosity curves over a wide range of temperatures," and also developed napalm, jellied gasoline. I briefly discuss here Division 8 involving explosives and Division 11 work by chemical engineers.

NDRC Division 8, Explosives

George Kistiakowsky of Harvard and Ralph Conner of Pennsylvania University were the authors of the chapters concerning Division 8. I mention some high points in the history of Division 8, mostly derived from W.A. Noyes' book.

> In the early days of the work on explosives the Service representatives had attitudes ranging from tolerance to definite antagonism . . . for the creation of an explosives program within NDRC. It is probable that in the large majority of cases lack of co-operation came from the sincere but smug conviction that the Services knew all that was to be known about explosives and that amateurs could not make significant contributions. [Noyes, page 20].

* * *

Twenty organic chemists met in the basement of Dr. Roger Adams home in 1940 to discuss what they might do with respect to the defense effort. RDX (also called cyclonite) had been recognized since the end of World War I to be an especially powerful explosive, but it was difficult to prepare, used scarce resources, and could be produced with only a low yield. Over the next several months, the organic chemists studied the literature on the subject, and embarked on a program to understand the complex reaction system. Step by step, they developed a synthesis that used abundant inexpensive reactants and gave a high yield of RDX, and they recycled a reaction co-product back into starting material. They replaced the previous batch method with a continuous process. Based on the organic chemists' results, there was a successful pilot plant by the end of 1941 and a successful large-scale factory by June of 1942. It was estimated that the NDRC work saved the government 130 million dollars in construction costs of the factory. With some die-hard exceptions, the Army and the Navy were enormously impressed, and they became willing to work with the civilian chemists and with NDRC.

Division 8 established a central Explosives Research Laboratory (ERL) near Pittsburgh, Pennsylvania, and later an Underwater Explosives

Research Laboratory (UERL) at Woods Hole, Massachusetts. It also placed contracts with university scientists and in industrial laboratories.

Top physicists, such as J. von Neuman and Hans Bethe, and top physical chemists, such as R.C. Tolman, J.G. Kirkwood, G.B. Kistiakowsky, and E.B. Wilson, Jr., took on the task of understanding in detail the mechanisms and effects of combustion, explosions, detonations, and shock waves in air, water, and solids. Comprehensive understanding of explosives came from top level theory, from tens of thousand small tests in the laboratory, and from numerous field tests on small and large scale. Propellants of widely different properties were required to propel shells from artillery, mortars, bazookas etc., and to provide jet propulsion of rockets. From their understanding of explosives, Division 8 scientists could calculate the magnitude and duration of the force of propulsion, which depended in a known way on the size and composition of explosive pellets. Pure RDX or pure TNT was not used, but combinations of explosives or of explosive combined with additives were used for various tasks. Underwater explosives differed from those used on land or in the atmosphere. The Armed Services brought large and small problems to NDRC and adopted for military service many of the solutions provided by NDRC Division 8 (Noyes, pages 15 to 139).

In the Caltech chemistry department, Linus Pauling and R.B. Corey had a project with sixty participants, who analyzed rocket propellants and other material submitted by the Army and Navy, developed new double-base powders, and performed numerous other services. Also there was a major war-research program operated by the physics department at Caltech concerning rocket propulsion:

> A project was developed on campus for the design and development of rocket projectiles. By September [1941] it was well underway, under the able leadership of physics professor Charles C. Lauritsen, and it continued through the deployment of the atomic bombs on Japan, until September, 1945. The project started in the Kellogg Radiation Laboratory with a federal grant of $200,000 and quickly grew to an $80 million industry, employing nearly five thousand workers at a number of nearby sites. In 1944, the one-millionth Caltech rocket, nicknamed "Holy Moses," rolled off an assembly line outside of Pasadena. Caltech had become, in the words of Lauritsen's assistant William A Fowler, "a branch of the Bureau of Ordnance."
>
> (Courtesy of Archives of California Institute of Technology, 2001).

Professor George Kistiakowsky of Harvard was the leading NDRC physical chemist on the development and use of explosives. In the middle of the war he transferred from NDRC Division 8 to the Manhattan Project. To detonate a nuclear bomb, a chemical explosive drives together two subcritical masses of uranium 235 or plutonium-239 in order to exceed the critical mass for nuclear chain reaction. Using his deep understanding of explosives, Kistiakowsky played a major role in designing the first nuclear bombs.

Division 11, Chemical Engineering

Division 11 of NDRC involved forty unrelated Service projects handled by seventy research and development contracts (Noyes, pages 343 to 430). One effort was the production of pure gaseous oxygen for use in submarines, high flying aircraft, hospitals, and elsewhere. Another line of work was to produce flares and photoflash bombs for the Navy, including underwater flash illumination. There were projects to develop corrosion-resistant paints and coatings, especially underwater coatings.

Early in 1941, the Air Force used several different hydraulic fluids, each of which worked over only a narrow range of temperature. At minus forty degrees Fahrenheit, hydraulically operated guns and gun turrets on aircraft locked up and would not work; and at moderately high temperature, the fluids became thin and leaked. By the summer of 1942, chemical engineers, using known engineering principles and a large amount of testing, produced a hydraulic fluid that had constant viscosity from minus forty to plus two hundred degrees Fahrenheit (–40 to +93 degrees Celsius). The armed Services quickly adopted this new hydraulic fluid, for use with heavy artillery as well as by aircraft.

* * *

The Chemical Warfare Service was responsible for flame throwers and incendiary bombs, and NDRC Division 11 led in the development of these weapons.

> When work on incendiaries was undertaken by NDRC in the summer of 1941 the large-scale use of incendiary bombs was a relatively new concept. The effectiveness of the incendiary as a weapon of attack had been shown during the blitz on England, where destruction per ton of incendiaries dropped was about five times as great as the damage from a ton of high-explosive bombs. [Noyes, page 388].

The German incendiary bombs used a kilogram of magnesium [2.2 pounds], but magnesium was in short supply. Starting in September 1941, NDRC set out to produce an incendiary bomb at least as effective as the

German magnesium bomb, but to avoid use of strategic materials. Projects were set up at Harvard, Brown, and Chicago universities and at chemical and oil industries. All of these projects sought to use thickened or "jellied" gasoline. Harvard produced the best product, which was named "napalm." During the war, eighty million pounds [36 million kg] were used in flame throwers and in incendiary bombs. Typical of NDRC projects, investigators made numerous field tests to achieve a complete understanding of the weapon:

> Several experimental lots of the gasoline-jelly bomb were used in testing work. Various mechanical improvements to increase the reliability of functioning were incorporated in the design as information was obtained from the testing work.
>
> In order to simulate the performance of the bomb when dropped from aircraft, a mortar was constructed at Standard Oil Development Company and used to fire the bombs downward onto a target at any chosen velocity. The mortar was mounted on a movable crane . . . so that the point of impact of the bomb on the structure below could be closely controlled. Various sections of typical enemy domestic and industrial structures were employed to test the penetration, functioning, and fire-raising characteristics of the 6-pound oil bomb. . . Over 20,000 bombs were tested by Standard Oil Development Company before the final design was standardized. [Noyes, page 391].

Dugway Proving Ground built twelve realistic two-family Japanese houses and six German houses, including furniture, and, as further tests, aircraft dropped incendiary bombs on these houses.

WAR VASTLY WORSE THAN NDRC EXPERIENCE

In twenty-seven attacks between January 6, 1945 and June 19, 1945, large fleets (57 to 520) of B-29 bombers struck thirteen Japanese cities with incendiary bombs, especially Tokyo. The aircraft dropped forty-eight thousand tons of fire bombs, destroyed 126 square miles [328 square km] of built-up areas (Noyes, pages 408, 409). The incendiary bombs in the cities set up fire storms, where hot air from buildings burning over a wide area rose up and sucked in surface air from all sides at hurricane velocities to achieve temperatures so high and winds so powerful that all structures were destroyed. The damage done (Noyes, page 409) was greater than that caused by the nuclear bomb that struck Hiroshima,[#] which had an explosive force of twenty thousand tons of TNT and destroyed four square miles [10 square km]. In both cases the attacks were against cities, and it is highly probable that most of those killed were civilians. This account is of history, and history includes many harsh facts.

The total number of people directly killed by World War II "is estimated to have been 55 million dead—25 million of those military and 30 million civilian . . . The Allied military and civilian losses were 44 million; those of the Axis, 11 million."[§]

AN NDRC BLUNDER

On pages 82 to 84, Noyes described NDRC work on underwater explosives on the ship *Reliance* operating out of Woods Hole Harbor, Massachusetts:

> This was the evaluation of underwater performance of loaded naval munitions . . . Steadily moving against the tide, the *Reliance* kept the main cable taut and thus insured controlled distances. The floats and weights maintained constant depth of immersion. After the shot the cable was reeled back with the attached gear . . . Those participating in the tests acquired only gradually the skill and experience as to how and when various gear must be cast overboard in such a way that the whole did not become a snarled mess. The candid snapshot **[Figure 7.1.]** indicates what happened when a mistake was made. On other occasions men were pulled overboard by loops of cable.

[#]"Hiroshima," Microsoft Encarta Encyclopedia. (c) 1993–1995, Microsoft Co., (c) Funk & Wagnalls Corporation.

[§]"World War II" *ibid.*

Figure 7.1. A failed experiment by an NDRC group studying underwater explosives. (Noyes, Page 83).

It would be interesting to know the full story of Figure 7.1 and of the people in it. The principal investigators and founders of NDRC were in their thirties or forties or older during World War II, and survivors today are in their nineties or over a hundred. The graduate students on those projects were typically in their early twenties then and in their early eighties now. If someone would contact the survivors in Division 8 to obtain their recollections, old letters, diaries, and pictures, one would probably find a richer vein of human interest and a more solid sample of scientific history than I have provided in this book.

CHAPTER 8

PRINCIPALS AND CONTRIBUTORS

"The purpose of this book is to present an almost forgotten history of secret war research. I focus on one narrow subject and a small number of individuals, but I try to give an in-depth study of these individuals and of what they did." Quoted from the Preface of this book. In the first six chapters, I recorded what I could learn about the actions of this small group of interesting people on these almost forgotten projects, and in this chapter I present later development concerning some of these people.

Principals in this Book.

ROSCOE GILKEY DICKINSON

June 10, 1894–July 13, 1945

Courtesy Archives, California Institute of Technology, Historical Files of Roscoe Dickinson,

by Linus Pauling, July 14, 1945.

Friends: We have come together today, the colleagues, students, other friends of Professor Dickinson—of Roscoe Gilkey Dickinson—here in the Gates and Crellin Laboratories of Chemistry where his life work was carried on, not, as we have done so often in the past, to discuss with him some detailed scientific experiment, but instead to recall the man himself—to think again about his life, his scientific work, his relation with his colleagues and his students, his influence on this institution in which he worked.

Roscoe was born in Maine fifty-one years ago. His father was a musician, his maternal grandfather a New England sea captain. From his father he inherited the great love of music which formed such an important part of his character—perhaps it was from his sea-captain grandfather that he got the spirit of roving inquiry that led him into science.

As an undergraduate at the Massachusetts Institute of Technology he studied chemical engineering, in which he received his bachelor's degree in 1915. But he was dissatisfied with the commercial aspects of chemical engineering, and strongly attracted to pure science, he came under the influence of Professor Arthur A. Noyes, under whose guidance he carried out his first scientific research, a determination of the vapor of water above salt hydrates. For two years he worked as research assistant at M.I.T., and then in 1917, at age of 23, he came, at the request of Dr. Noyes, to this Institute in Pasadena, and began the period of twenty-eight years of devoted service to the Institute.

The younger men among you may not know how great was his contribution to the development of the work in chemistry at the California Institute of Technology, and especially of graduate study and research in chemistry. He himself was the author of about half of the first dozen or so papers published from the Institute and he lived to see over one thousand papers published from the Gates and Crellin Laboratories. In 1920, after three years in Pasadena as Instructor, he received his degree of Doctor of Philosophy—this was not only the first Ph.D. in Chemistry, but the first doctor's degree ever granted by the California Institute. Not only directly, but also through his intimate friendship with Dr. Noyes, who relied upon his honest judgment and devotion to science, he assisted greatly in determining the nature and upholding the standards of advanced work in chemistry. This work he continued through the years of service as Chairman of the Division Committee on Graduate study and Research, and later as Acting Dean of the Graduate School of the Institute.

But the greatest service of all was his direct influence on the many young men who carried on their advanced study with him, and who learned from him how to carry through a logical argument, how to distinguish between a scientific proof and a surmise, how to be honest in appraising the results of an experiment. I remember the first day when I first saw Roscoe! I was a fresh young graduate student, who had been waiting his return from a summer in Massachusetts to direct my research; he came down the basement hall in the Gates Laboratory, with shining eyes, walking eagerly, almost jauntily, glad to be back in his laboratory, where he could carry through the searching study of the secrets of nature by his x-rays, supply the methods of rigorous analysis which he had developed to elucidate the structure of the crystals which for generations had posed puzzling questions for past chemists. And he explained to me the need for honest, careful, sound, analysis in scientific work, as in his later years he explained his scientific methods to other students, a score of them, and he helped them find the right attitude of the scientific man toward his work.

The field of structure of complex crystals which occupied Roscoe's attention during his first decade of his scientific activity and comprised the content of his first dozen papers was an ideal one for him, permitting him to employ fully his gift for clear thinking and logical analysis. When he began work the structures of only a few simple crystals were known, and the nature of the quantitative relations governing the interaction of x-rays and crystals was sufficiently doubtful to make an attack on complex crystals—the chlorostannates, chloropallidates, complex cyanides, tin tetraiodide—seem hardly worth while. Roscoe set himself the problem of developing a method of attack which would be rigorous and completely reliable but which would depend only on the known qualitative properties of x-rays; the method which he developed was completely reliable, and he had complete success in it application, not only to complex inorganic crystals, but also to the first organic crystal, hexamethylene tetramine, for which a crystal-structure determination was never made. A few months ago, in summarizing before an advanced class the work of early structure investigators, I pointed out that Professor Dickinson is the only one whose work has been found to be completely free from error; and it is characteristic of him, of his inherent modesty, that in every case his work has been found to be much more accurate than he claimed it to be.

In both his teaching and research he ranged over a broad field. Over the years he taught general chemistry, chemical thermodynamics, statistical mechanics, photochemistry, reaction kinetics, and other advanced courses, always with his own thorough and sound treatment. In research, in addition to crystal structure, he studied the mechanism of photochemical and thermal reactions, the Raman spectra of dissolved ions and of gases (as one of the first in this field), the properties of neutrons, the chemical properties of Vitamin C—each new problem he attacked with ever-youthful enthusiasm, always from the motive of pure scientific curiosity, the desire to know the answer, and always in the spirit of honest, critical skepticism. The only part of chemistry he did not work in was that of proteins and similar natural products. He made one try—to prepare crystalline pepsin for denaturation study—and gave up the work as far too unprecise for him.

And so the years went by, and he was happy with his work, in his home, his children, his music, his trips to the desert and his painting—until a few years ago, the war brought a change in the world and in him, He was asked to give up the work in pure science which he had loved—which over the years he had refused to neglect for even a few days when he was asked by industry to serve as a consultant—in order that he might be of service to the Nation, helping, as part of the Office of Scientific Research and Development, to solve problems confronting the armed forces in time of war. For the last three years all his efforts were devoted to these war problems. He worked intimately with some of you young men, found the answers to the questions put to him, developed valuable apparatus, traveled across the continent and to the tropics—and then, suddenly and without warning, his life was brought to an end by the insidious but violent attack by an enemy against whom man has as yet not succeeded in developing a defense.

Today, as we think back over the life of our friend Roscoe Dickinson, we say, I feel, find much to comfort us. His life was not a long one, but it was a full one and a happy one and, because of his complete devotion to his purpose, a most effective one. He was happy in his home, which he with his wife designed and built, with his children, now mature, and with his granddaughter, with his music, with his desert trips, and his paintings; he was happy and contented over thirty years with his teaching and scientific research and his contributions to the life of the Institute and during the past few years he was proud that he could be of service to the Nation by his war work, and in addition he enjoyed keenly the new experiences that this work brought to him—the long airplane trips, the life in the tropics, the close contact with people of a different sort from those of the laboratory, and especially the intimate contact with his boys, the young scientists whom he loved and who loved him.

DON M. YOST
1893–1977

"Tribute by Terry Cole," ENGINEERING AND SCIENCE, Vol. 41, No. 128–29 (1977), continued from Chapter two.

Even after his active participation in research declined in later years, his interest in scholarly matters continued and was expressed through correspondence with colleagues, students, and members of the Iron Nail Club—La Sociedad des Clavos Hierros Cuadrillados, an intellectual and philosophical corresponding society founded by Don (Cisco) and Pancho P. Gomez of Idaho City, Idaho, dedicated to the free though intermittent discourse on politics, art, science and humor; listing (by noms de plume only) many of the great and near great of American science and free enterprise. He also wrote on mathematics, the historical aspects of science, and—most memorably—book reviews.

Don's book reviews, published in the *Journal of the American Chemical Society* and *Nuclear Science and Engineering*, have become minor classics of their genre, filled with his perception, erudition, and wit. As a brief exemplar of his style, he began the review of the volume, *Applications of Nuclear Physics:* "There was a time when those of us born west of Dodge City pictured England as a pleasant, provincial island where the men raced around the countryside in Rolls Royces chasing small foxes, where the women rode through the streets on horseback protesting oppressive taxes, and where millions of innocent children were brought up on Latin, *Alice in Wonderland,* W. Shakespeare, and on the exploits of the privateer Sir Francis Drake. But this picture is, in part, now notably different, the change really having been initiated by a transplanted New Zealander (Rutherford) and a visiting Dane (Bohr)."

Characteristically independent, he was always a staunch defender of individual independence against the strictures of official policy. His normally gentle wit became a rapier when deflating administrative pomposity or bureaucratic presumption. One of his former students has called him "The foremost anti-stuffed-shirt in American science."

Don prized and encouraged originality and independence in his students. He expected them to take the initiative in getting the work done; yet when genuine problems arose, he was always generous with his time in discussion and in sharing his vast scientific experience. Caltech can he a rather intimidating place for a new graduate student realizing how many scientific giants inhabit this small campus and how much he has to learn. Don's courtesy, informality, unfailing good humor, and grace in instruction toward this former student are memories I shall always treasure.

Terry Cole, PhD '58, who is a senior staff scientist at the Ford Motor Company Research Staff, spent most of the last academic year back at Caltech—this time as a Sherman Fairchild Distinguished Scholar.

ROBERT MILLS
1922–1991
by Richard LeSar

There are a few things I want to tell you before I begin to talk about Bob. Bob did his thesis in 1950 at Stanford University with Harold Johnston. A few years later, Johnston's group was joined by a rather bright young undergraduate, who Johnston arranged to spend a summer at Los Alamos. Bob and Rene took this young man under their wings. He moved on to graduate school at Harvard and went on to have a pretty successful career, becoming a professor at Harvard and receiving a Nobel Prize. As many of you know, that young man was Dudley Herschbach. Many years later, another young man went to Harvard as a graduate student and worked with Dudley. In 1978, Dudley packed this student off to Los Alamos for a summer with Bob. As you might guess, I was that student.

When I wrote the acknowledgments to my thesis in 1981, I spoke of my time in Los Alamos, and said: "The summer I spent in Los Alamos was educational and broadening. My greatest thanks go to Dr. Robert Mills, who acted as my mentor and friend."

That sums up pretty well my relationship with Bob. He was a friend and I loved him dearly, but he was also much more; he was a mentor, a teacher, and an important influence on my life as a scientist and, I hope, as a person.

I want to deal with what may at first seem to be an impersonal subject, namely Bob as a scientist. However, while a person is not defined by what he does, he is by how he does it. A couple of days ago, I found a quote by Hans Mohr that describes good science. When I read this quote, I was sure that he was describing Bob.

> Be honest; never manipulate data; be precise; be fair with regard to priority of ideas; be without bias with regard to data and ideas of your rival; do not make compromises in trying to solve a problem.

Bob is one of the very few (if not the only) scientists who lived up to all of these. I wish I did.

One of Bob's most endearing qualities is that he didn't care about credit. He never worried about who was first author of a paper or if people properly referred to his work. While I think he was proud of being made a Fellow of the Laboratory, I had the feeling that he thought the whole thing was really rather unimportant. I wish I had the self-confidence to be as unmoved by recognition as he was. Indeed, he was great in bringing me down to earth when this brash young Harvard graduate got carried away. As in all things he did, he did this gently, with a wry comment.

Bob was a terrific writer. David Schiferl and I were ecstatic when Bob read over a paper we had written and had no suggestions. Usually, Bob found many errors and inconsistencies and would work until it was as perfect as it could be. It is not that he was a hard "taskmaster," it is that he set such high standards, especially for himself, that we felt compelled to meet them.

Not only did Bob set a terrific example of how to do science, he was fun to be with. In his quiet way, he was an incredibly funny man. The mangled metaphors and fractured cliches that he collected were priceless. Many of them I won't share with you in public, but an example would be "He is a land mine of information" or "I am waiting at the station until the gravy train arrives."

To sum it up, Bob was a great scientist and teacher. His success, though, came from his qualities as a person; his honesty, especially with himself, and his integrity.

JOHN W. OTVOS

Present Address: 110 Ravenhill Road, Orinda, California 94563, Phone (510) 254-1986

Education: 1939 B.S. Harvard University, summa cum laude (Biochemistry) 1943 Ph.D. California Institute of Technology (Physical Chemistry)

Position: 1976– Staff Senior Scientist, Lawrence Berkeley Laboratory, University of California, Berkeley. Co-Principal Investigator in research on solar energy storage and photochemical storage systems modeled after photosynthesis, and the production of liquid fuels from green plants.

Past Positions: 1941–1945. Caltech project under National Defense Research Committee,. Analytical methods development and testing of chemical warfare agents, including jungle field tests in Florida and Panama.

1946–1976. Shell Development Company, Emeryville, California. Research in radiochemistry, spectroscopy, polymer and surface chemistry, petroleum chemistry and analytical instrumentation.

1946–1951. Carried out research using isotopic tracers on the kinetics and mechanisms of hydrocarbon reactions: hydrogenation, hydrogen exchange, isomerization, alkylation, thermal cracking. As Shell's first radiochemist, began program of applying radioisotopes to instrumentation, chemical analysis, and tracer studies in laboratories and manufacturing plants.

1951–1952. Training assignment as technologist at Shell's Wood River, Illinois refinery.

1951–1957. Assistant Manager of Spectroscopic Department at Emeryville. Responsible, together with Manager, for the research and analytical service of a group of 30 people in mass spectroscopy, emission spectroscopy, ultraviolet, visible and infrared spectrometry.

1957–1971. Manager of Chemical Physics Department. Responsible for planning and execution of basic research program involving staff of 27 with 16 professionals, whose research fields included polymer physics, surface chemistry, molecular spectroscopy, magnetic resonance, quantum chemistry, radiochemistry, and radiation chemistry.

1971–1976. Manager, Analytical Research in Shell Development's Analytical Department of 175 people. Moved to Houston, Texas in 1972 with the consolidation of Shell's research activities there. Responsible for program generation, technical excellence and relevance of all broadly based analytical research. Accountable for annual budget of $1,300,000 (in 1975 dollars). Research spanned the entire analytical field: Separations, molecular spectroscopy, physical measurements, organic analysis, electrochemistry, elemental analysis, etc. Initiated programs in a number of new techniques since 1970: ESCA, laser-Raman spectroscopy, ^{13}C NMR, plasma emission spectroscopy, energy dispersive X-ray fluorescence.

Partial autobiography of John Otvos

My father, Otvos Adorjan, (Jani) was born in Budapest on October 11, 1890. When we immigrated, he changed the "Adorjan" to "A. Dorian" to make it easier to pronounce. Whenever he was asked what the "A" stands for, he could truthfully say it doesn't stand for anything. His mother died in childbirth and he was raised by his father and stepmother, who later also had a son, my father's half-brother.

My mother, Lovinger Ilona, (Ilonka) was also born in Budapest, but the date is not quite certain. This is because she had a fetish for hiding her real age and she didn't tell consistent stories to all her friends. She was an only child, so on her side, too, I had no full-blooded first cousins. My closest relatives through her were her first cousins, Paula, Ilona, Anna, Erzsi, Archibald, and Manzi Schwartz ... Their aunt, Anna, was killed in Auschwitz.

My mother's home was at Damjanics utca 32; it was an apartment house with an interior courtyard. The apartments must have been pretty big because my mother first lived there with her parents, Adolph and Maria, and two domestics, Annaneni and Zsuzsi. Then my father moved in after the marriage on August 10, 1916. So after I was born [1917], there were 7 of us living there, but I don't remember ever hearing that we were in cramped quarters. Annaneni arrived on the scene from the town of Gyomro' in the country in about 1903 when she was 23 and my mother was 9. She became my mother's governess as well as the family cook.

My parents met in 1911 at a theater performance of a play called, "Testor" (Bodyguard). About three years later my father graduated from the Royal Music Academy where he was a classmate with Eugene Ormandy. He was an accomplished pianist and composer, and the Academy was quite prestigious, having produced many world-famous musicians in that decade, some of whom later showed up in America. My father was very talented and blessed with absolute pitch. To make a living after his marriage my father worked as a clerk at the electric utility company. But he kept up his music on the side and composed quite a large number of popular melodies right up to our emigration. A few became big hits all over Europe and had their lyrics translated into four or five languages. My mother was an excellent musician in her own right. She taught piano to augment the family income in the early days in New York, and maybe also when we were still in Budapest.

It was my father's music that supported the family through most of the first decade in America. Neither of my parents could speak English when we arrived, although my mother had studied it for a short time in school. But my father's music was readily salable. The day to day earnings I think came mostly from playing the piano in the "pit" in movie theaters during silent pictures. Then he gradually became associated with Broadway where he played in the orchestras for musicals and wrote some of the pieces for musical revues.

Central Europe after the Great War was not a pleasant place to live. This was particularly true in the defeated countries like Austria, Hungary, and Germany. There was poverty and political disorder—two revolutions in Hungary a year apart, one communist and the other anti-communist. There was runaway inflation, and the mechanism of distribution of goods, especially food, had broken down. Besides all this, for the Jews there was the undercurrent of anti-semitism, which always gets worse when times are bad. I think this was when I first became a factor in my parents' plans. They wanted something better to look forward to and they needed more hope for healthy conditions for raising their child.

But our emigration would never have happened without the help of the Black (Schwartz) brothers. There were four of them: George was the oldest, then Maurice and John, and finally Bela. As I said, although my mother called them her "cousins", that wasn't strictly true. They were somewhat older, and evidently little Ilonka was the apple of their eye. Just before the war as young men they emigrated to the United States. They all worked hard in the clothing business, saved their money, and by 1921 were quite well-to-do and able to send for Ilonka and her family, pay for their passage, and set them up in a small apartment in New York City.

In 1923 I started school at P.S. 10, a few blocks East of Morningside Avenue at about 118th Street. My parents in the meantime were worried about the school's quality, and one of their American friends told them that Johnny was so smart he should go to the best. He recommended the Ethical Culture School, run by the Ethical Culture Society, which was founded by Felix Adler in the late 19th century. Its guiding principle was that each person has some unique attribute that should be nurtured. Classes were small and the teachers were excellent. (There were 77 in my senior class. Of the 36 boys 10 were accepted at Harvard.) But the school was expensive, way out of our reach. My father must have written one of his persuasive letters because they gave me a full scholarship, which I maintained over the whole 11 years I was there.

It must have been in the fourth or fifth grade that my parents gave me my first chemistry set. I never turned back. Friends had them too and we played with them and read about chemistry together. Several of us made copper-plated jacks and sold them to classmates.

Most of the student body of Ethical Culture, probably more than 85%, was Jewish and quite wealthy. This gave my parents no problem, but in retrospect it seems ironic. We had left Hungary largely because of anti-semitism there and to start over in the New World where we could get "assimilated." Now all my friends were Jewish and so were most of my father's associates in show business.

Society in the 20's was almost totally segregated. When no country clubs would admit Jews, the Jews built their own where everyone was Jewish. Neither kind ever admitted Negroes. It was almost impossible for a Jew to get into medical school, or any professional school for that matter. It's well documented that even in the early 40's Richard Feynman was turned down by Berkeley's physics department because of Chairman Birge's extreme prejudice.

A few years later three big changes happened in our lives. My father got a job in the movies at Vitaphone Corporation, a Warner Brothers subsidiary, we moved from Morningside Avenue to Riverside Drive, and I graduated from the Ethical Culture lower school of 6 grades to the upper school called Fieldston.

I remember that we had happy years on Riverside Drive. The Vitaphone job was steady so the stock market crash in October of 1929 had absolutely no effect on us. My father never owned a stock, and even if he had, he never would have bought on margin. But movies still went on during the depression and Vitaphone produced a short subject every week, which was written by my father and collaborators. Although Vitaphone paid well, he never liked the grind of mass production and he had a low opinion of the literary talents of his boss. Once, in frustration, he told the boss he couldn't produce a decent script with such strict time limits and that if he had an extra day or so he could do much better. The reply became another family saying. "I don't need it better, I need it Tuesday," said the boss.

I was getting pressure at school from my science teacher, Mr. Klock, who thought I should have the best possible science education, which he said was at Harvard. I applied also to Yale and Johns Hopkins, which had a high regard for the Fieldston School and served as my fall-back choice. At all three colleges I also applied for a scholarship. When the replies were in, I had been accepted everywhere. Johns Hopkins offered me a full tuition scholarship of $400 for the first year, Harvard … awarded me a half-scholarship of $200, and Yale offered no scholarship. Harvard it would be. In September of 1935 Allan Kalmus and I left by train for Cambridge. We roomed together in Weld Hall in the Yard. It was a good location, near Mass. Avenue, Widener Library, and the Freshman dining hall. Supposedly there was open seating there, but we found that a number of tables were only for the elite, the men from Exeter, Andover, and Groton. If we sat at one of them nobody would talk to us.

Early in 1937 there was another big change in our lives. My parents moved to California. My father was lured by the movies and he had many friends in Hollywood, both from Broadway and from the Hungarian community. As I recall, he took a trip West by himself in 1936 to have some interviews. The trip resulted in negotiations with Universal Pictures for a screen writer contract. The contract was the standard kind. The salary was very good but the term was only one year with option to renew. My parents headed out in January while I was in college. When the year ended, Universal didn't exercise their option and my father was unemployed. He never held another steady job. That was the nature of Hollywood—always prospects, always promises but seldom a commitment. From time to time there would be a rewrite assignment for a couple of weeks, or a short radio script to do, but mostly it was waiting for the phone to ring. It was a terrible way to live—perpetual anxiety, and not very good for your blood pressure either.

A small inheritance came to my mother from Maurice. That was when my parents moved to Hollywood at 1351 North Ogden Drive. It was a comfortable home with a nice garden. They moved in early in 1939 before I came home from college after graduation. It was the last home my parents would live in.

In June of 1939 I graduated from Harvard with a B.S. in biochemistry. Moreover I graduated summa cum laude, largely on the basis of the research I did Senior year with Dr. John T. Edsall at the Harvard Medical School. Only about one percent of the graduating class received those honors. That was a great source of pride for my parents.

To make ends meet, my parents decided to open a book shop because they did have some expertise in that area. It was called the "Folio Book Shop" and was on Santa Monica Boulevard in Beverly Hills. Mostly Folio dealt in conventional and current books. The shop was small and quiet. But the shop served its purpose and supported them both till they died.

About the beginning of 1941, as a graduate student at Cal Tech, I started working for the NDRC (National Defense Research Committee) on a classified war project related to poison gas warfare. Our group was directed by Roscoe Dickinson, with whom I later got my (Ph.D.) degree. Other sections were led by Don Yost at Cal Tech, Wendell Latimer in Berkeley, and Malcolm Dole at Northwestern. Later on, in 1944, we all joined up for our stint in Panama, where we worked as civilians with the Army doing field tests on an uninhabited island. When our job was done, I came home (by airplane) with a detour to New York so I could attend Allan Kalmus's wedding on September 9 and have a short holiday with my "uncles."

When I started working at Cal Tech again, I promptly met Margie. She was doing classified work at the other end of the basement hall. Her friend, Bettie Curland, arranged a blind date for us after the Caltech Chemists baseball game that I was playing in on the evening of October 9. It wasn't long afterward that I introduced Margie to my family and some of their friends. Of course she also met Annaneni then, who had cooked the dinner as usual. Everyone was very polite to Margie and tried, without much success, to keep from lapsing into Hungarian in her presence. But some of my mother's friends were slow to accept her. They wondered why Jancsi couldn't get interested in a nice, Jewish girl.

Our wedding was held in Margie's house, 1701 Meridian in South Pasadena, and we thought it would be appropriate to have my father play the wedding march on the piano. The Unitarian ceremony was to be very simple and attended only by family, with the exception, of course, of Bob and Rene Mills because Bob was best man.

With the war over, the recruiters came on campus. I was invited to visit Shell Development in September, so Margie and I came up again for a few days for the interview. We had pretty well made up our my minds that Shell was it, but U.S. Rubber in New Jersey as well as Hercules in Delaware also offered me visits. But it was a fine trip—the Superchief to Chicago and then the 20th Century Limited to New York. U.S. Rubber really went all out to get me. We were wined and dined, and then on the second day they asked me what Shell had offered me. I told them $375. a month. The interviewers went into a huddle and came back cheerfully saying that they were able to up that to $385!

We left for Berkeley on New Year's day of 1946. Mrs. Ainsworth of the Shell personnel department found us an apartment on the second floor at 1136 D Hearst Avenue, a block from University and San Pablo. About a year later when Margie was expecting, we were able to move to a ground floor unit in the same complex at 1140 A.

Rene Scott Mills, 1945.

Bill Shand, 1944, Caltech, NDRC. Photographer of Chapter 5.

Figure 8.1. Photographs courtesy of Rene Mills, and unknown Army photographer.

JOHN R. THOMAS
1921–2001

Birth Date:	August 26, 1921—Anchorage, Kentucky	
Education:	University of California—B.S. Chemistry	1943
	University of California—Ph.D. Chemistry	1947
Experience:	National Defense Research Committee (NDRC)	1943–1944
	Manhattan Project	1944–1946
	General Electric	1947–1948
	Chevron Research	1948–1949
	U.S. Atomic Energy Comm.	1949–1951
	Chevron	1951–1986
Positions:	Assistant Chief Chemistry Branch, USAEC	1949–1951
	Research Scientist, Chevron Research	1951–1968
	Manager, Ortho R & D, Chevron Chemical	1968–1969

	Assistant Secretary, Chevron Corporation	1969–1970
	President, Chevron Research Company	1970–1986
	Vice-President, Chevron Corporation	1983–1986
	Director, Cetus Corp.	1978–1986
	Compensation Committee, Cetus Corp.	1981–1986
	Chairman of Board and Director, GA Technologies	1984–1986
	Director, Coordinating Research Council	1980–1986
Memberships:	ACS, AAAS, SAE, Commonwealth Club, Sigma Xi	

John R. Thomas – Died May 6, 2001 at the age of 79. A retired President of Chevron Research Company and Vice President of Chevron Corporation. He received both his B.S. and Ph.D. degrees from Cal Berkeley, and continued his affiliation with the University as a Trustee of the Berkeley Foundation and as a Berkeley Fellow. During World War II, he worked on the Manhattan Project and later became Assistant Director of Research for the Atomic Energy Commission in Washington, D.C. for five years before returning West to Chevron as a research chemist. After retirement, he became an activist for environmental, population, and immigration concerns. A mountain man, he backpacked, skied, hiked, and shared the serenity and beauty of the Sierras with friends and family at his mountain cabin. He is survived by his best friend and wife of fifty-seven years, Mitzie, son Richard and daughter Jonnie Jacobs, both of Piedmont; grandchildren Elizabeth and Alex Thomas Matthew and David Jacobs; and brother Robert and family of Annapolis MD.

ANDREW A. BENSON

Born September 24, 1917 in Modesto, California

Education

University of California, Berkeley, College of Chemistry 1935–1939. B.S., 1939.

California Institute of Technology, 1940–1942. Ph.D., May 1942. Thesis: I. Synthesis of Fluorinated Analogs of Thyroxine. II. Oxidative Degradation of Sphingosine Analogs. III. Inhibition in the Slow Muscle of Pecten.

Professional Experience

1942–1943	Instructor, Department of Chemistry, University of California, Berkeley. C11 and C14 research on photosynthesis with Sam Ruben and Martin D. Kamen. Isolation of dark CO2 fixation products of photosynthesis.

Figure 8.2. Ruth Carkeek and Andy Benson, married May 22, 1942. Photograph supplied by Andy Benson.

1944–1945	Civilian Public Service, Aerial Photogrammetry, U.S. Forest Service;
1945–1946	Research Associate, Stanford University. Antimalarial drug synthesis with F.W. Bergstrom.
1946–1954	Research Associate and Assistant Director, Bio-Organic Group, Radiation Laboratory, University of California, Berkeley. Research on the Path of Carbon in Photosynthesis with Melvin Calvin (Nobel Laureate, 1961).
1955–1961	Associate Professor, Professor of Agricultural and Biological Chemistry, The Pennsylvania State University. Research on lipid metabolism, neutron activation analysis, sulfocarbohydrate metabolism. Discovery of phosphatidylglycerol and the plant sulfolipid.
1961–1962	Professor-in-Residence, Laboratory of Nuclear Medicine and Radiation Biology, Departments of Biophysics and of Physiological Chemistry, School of Medicine, University of California, Los Angeles.
1962–1989	Professor of Biology, Scripps Institution of Oceanography, University of California, San Diego. Research on lipid

metabolism, recognition of wax ester as a major marine metabolic energy source. Recognition of coral mucus lipid as an important medium for energy transfer in the coral reef ecosystem; studies of mangrove physiology and metabolism; study of calcitonin regulation of calcium transport in the spawning Pacific salmon; discovery of the non-toxic metabolic intermediates of arsenic metabolism in algae.

1989– Professor emeritus. Discovery of growth stimulating effects of methanol on algae and higher plants.

Memberships

1973 National Academy of Sciences.

1981 American Academy of Arts and Sciences.

1984 Royal Norwegian Society of Sciences and Letters.

1992 Institute of Marine Biology, Russian Academy of Sciences, Vladivostok.

A portion of Awards and Honors

1950 Sugar Research Foundation Award, co-recipient with Melvin Calvin. For elucidation of the sequence of intermediates for sucrose synthesis in plants.

1962 Ernest Orlando Lawrence Memorial Award of the Atomic Energy Commission, For development of radiotracer methods in biology.

1972 Stephen Hales Award, American Society of Plant Physiologists.

1995 RIKEN Distinguished Scientist.

Honorary Doctor Degrees

1965 University of Oslo, Norway. For developments in lipid biochemistry and photosynthesis.

1986 Universite Pierre et Marie Curie, Paris.

BENSON'S CONTRIBUTIONS TO PHOTOSYNTHESIS

1. Used the first available quantity of radioactive carbon, C-14, for study of the path of carbon in photosynthesis with Melvin Calvin after the end of the war. Developed the technique of identification by solvent partition for identification of radiolabeled products of photosynthesis.
2. Discovered and identified the first product of photosynthesis, which over the world is produced at the rate of ten billion tons per year (Phosphoglyceric acid, with M. Calvin).
3. Discovered sedoheptulose and ribulose phosphates in the intermediates of the photosynthetic carbon reductive cycle. Discovered and identified

the compound which absorbs carbon dioxide from the air to produce all the compounds of plants and many bacteria (ribulose diphosphate). Recognized the role of ribulose diphosphate as carbon dioxide acceptor in the carbon dioxide reduction cycle of photosynthesis (with M. Calvin). Discovered and recognized the intermediates in the carbon reduction cycle of photosynthesis and the subsequent production of sucrose, the 'energy currency of plants'. Demonstrated the enzymatic catalysis of carbon dioxide fixation, recognized its identity with "Fraction 1 Protein", the main protein of leaves, and (with Jacques Mayaudon) was the first to isolate it as the enzyme, now called "Rubisco".

ARTHUR BECK PARDEE 1921–

Figure 8.3. Mr. and Mrs. Arthur Pardee, 1947.
Photograph supplied by Arthur Pardee.

Higher Education

University of Calif., Berkeley B.S. 1942, M.S., 1943
California Institute of Technology Ph.D., 1947
University of Wisconsin, Madison Postdoctoral Fellow, 1947–1949

Employment

1949–1961	Instructor, Assistant Professor, Associate Professor, Department of Biochemistry, University of California, Berkeley;
1957–1958	Senior Postdoctoral Fellow, Pasteur Institute, France
1961–1975;	Professor of Biochemical Sciences, Princeton University
1972–1973	American Cancer Society Scholar, Imperial Cancer Research Fund Laboratories, London
1975–Present	Professor of Biological Chemistry and Molecular Pharmacology, Dana-Farber Cancer Institute and Harvard Medical School and Chief, Division Cell Growth and Regulation, Dana-Farber Cancer Institute D810 44 Binney Street, Boston, MA 02115

The life-time theme of Professor Arthur Pardee's research is to discover and investigate fundamental mechanisms that control—or are deranged—by disease, especially cancer.

Arthur Pardee and I were graduate students together at Caltech in the 1940's, we worked with an army group that shot off poison-gas bombs in the swamps of Florida in 1943–45, and we were room mates at Caltech for about a year after the end of the war. I have followed his career with interest and have been much pleased by the glowing reports I have heard about his work during the years.

Memberships

American Academy of Arts and Sciences	1963
National Academy Sciences	1968
Institute of Medicine	1974
American Philosophical Society	2001

A portion of Awards and Honors

Paul Lewis Award (American Chemical Society)	1960
Honorary member Japanese Biochemical Society	1965
Sir H.R. Krebs Medal	1973
Rosensteil Medal	1975
Princess Takamatsu Award (Japan)	1990
Docteur Honoris Causa Paris	1993
Baehringer Manheim Award	1998
Distinguished Alumni Award, California Institute of Technology	1998

A major discovery from Pardee's research on nucleic acid precursor metabolism is feedback inhibition of enzyme activity, a fundamental control mechanism in animals as well as micro-organisms. The discovery of feedback inhibition led to his original demonstration of regulatory sites different from substrate binding sites on enzymes, a cornerstone of the allosteric generalization. His third major contribution, with Jacob and Monod, is regulation of gene expression by repression. A fourth is that transmembrane transport mechanisms are controlled by repression, followed by isolation of the first active transport component, the sulfate binding.

Figure 8.4. Dr. George Ruben.
Photograph supplied by George Ruben.

GEORGE C. RUBEN

Dept of Biological Sciences, Dartmouth College, Hanover, NH 03755
tel. (603) 646-2144, fax (603) 646-1347,

e-mail George.C.Ruben@Dartmouth.edu
Home
10 Shaw Street
Lebanon, NH 03766
tel. (603) 448-1726

Birth Berkeley, California; 4/29/41

Education

1963, BS Chemistry, University of California, Davis, Summa Cum Laude.
1972, Ph.D. in Physical Chemistry, University of California, Berkeley.
Postdoctoral Biochemistry and Applied Physics, Cornell, 1/71–6/76.

Faculty Appointments Research Associate I, Section of Biochemistry, Molecular and Cell Biology, School of Applied and Engineering Physics, Cornell University, 7/75–7/77, Dartmouth College, Hanover, NH: Research Assistant Professor of Pathology, 7/82–7/84; Research Associate Professor Biology, 7/84–6/94; Research Professor Biology, 6/94–.

Other

United States Patent 4,755,494 and 4,912,069 "Use of pectin and pectin-like material in water based ceramics" 7/5/88 & 3/27/1990; Pectin acts as a binder as well as slows clay drying. The pectin conducts water from thicker clay regions to thinner clay regions evening the drying process which makes it possible to again construct ceramic figures as large as the soldiers and horses produced by the Quin dynasty in China, 2000 yrs ago.

Orthopaedic prosthesis: Mechanism of wear of Ultra-high molecular weight polyethylene currently used in total hip and knee joint implants—see *J. Mater. Sci.* **28** 1045–1058 (1993).

Consulting—Industry

1985–86	IBM, Watson Research Laboratory, Yorktown Heights, NY
1984–87	Philip Morris's Research Laboratory, Richmond, VA
1986	Poloroid Corp., Cambridge, MA
1988	Polymer Technology, Wilmington, MA
1990–94	Lawrence Livermore National Laboratory, Livermore, CA
1992–93	YTC America, Camarillo, CA 93012
1992–97	Dow Corning, Midland, Michigan 48686

Professional Society Memberships

American Association for the Advancement of Science
American Chemical Society; Cellulose, paper and textile division
American Society for Cell Biology

Materials Research Society; Sol gels, xerogels and Aerogel ceramics
Microscopy Society of America
Royal Microscopical Society
Society for Neurosciences; Neuropathology

A portion of Fellowships and Awards

Sigma Xi, Phi Kappa Phi, Phi Beta Kappa
First Prize in Tribology TEM Photo Contest (1989)
Chemical Engineering News Award of Merit (1963)
Football Conference Honors
National Photo Contest Award (1970, 1972, 1974)
Mallory Cup Sailing Finals (1969)

Neuropathology: Alzheimer Neurofibrillary tangle structure, Paired helical filament substructure, and the structure of the microtubule associated protein tau. **Cell Biology:** extracellular matrix in plant cell wall with focus on cellulose and pectin structures, basement membrane structures, ie collagen IV and laminin network structure, toad red blood cell membrane cytoskeleton structure, Condensed DNA structure and DNA binding proteins, Photosynthesis, **Structure of noncrystalline materials** sol-gels; silica xerogels and aerogel structure and resorcinol-formaldehyde aerogel structure

High Resolution Electron Microscopy at the Molecular level (specimen preparation methods in general, freeze-etching and freeze-drying, vertical Pt-C replication, negative staining, stereoscopic imaging), circular dichroism, visible and fluorescent spectroscopy, low temperature specimen preparation techniques (cryomicroscopy).

HAROLD JOHNSTON

Biographical Notes
1920–

Professor Emeritus of Chemistry
University of California
Berkeley, CA 94720

Home: 132 Highland Blvd.
Berkeley, CA 94708-1023
PHONE: (510) 525-6810
FAX: (510) 528-6019

Born Woodstock, Georgia, October 11, 1920.

Education

Emory University, Atlanta, Georgia, 1937–1941, A.B. in Chemistry and English.

California Institute of Technology, 1941–42 and 1945–47, Ph.D. in Chemistry and Physics.

Employment

Civilian employee of National Defense Research Committee at Caltech, 1942–43, and field work with Chemical Warfare Service, Bushnell, Florida, 1944–45.

Instructor to Associate Professor of Chemistry, Stanford University (1947–56).

Associate Professor, California Institute of Technology (1956–57).

Professor of Chemistry at the University of California, Berkeley (1957–1991).

Dean of the College Chemistry (1966–1970).

Professor Emeritus of Chemistry since 1991.

Experimental research interests have been in the field of gas-phase chemical kinetics and photochemistry, involving oxides of nitrogen, ozone, fluorine, chlorine oxides, HO and HOO free radicals, etc., and in the fields of urban air pollution (1950–70) and stratospheric ozone (1971–).

Memberships

American Chemical Society (ACS)
American Physical Society (APS)
American Geophysical Union (AGU)
1965 National Academy of Sciences (NAS)
1972 American Academy of Arts and Sciences

A portion of Awards

1993 National Academy of Science Award for Chemistry in Service to Society.
1983 Tyler World Prize for Environmental Achievement.
1974 Pollution Control Award of the American Chemical Society.

A portion of Honors

2001 Festschrift, The Journal of Physical Chemistry, Volume 105, March
1997 National Medal of Science,
1991 Berkeley Citation, University of California
1985 Distinguished Alumni Award, California Institute of Technology.
1982 Certification of Commendation, Federal Aviation Administration.
1956 Gold Medal Award of the California Section of ACS

Honorary Doctor Degrees

1998 Honorary D.Sc., Northwestern University, Evanston, Illinois.
1965 Honorary D.Sc., Emory University, Atlanta, GA.

Figure 8.5. Mary Ella Stay and Harold Johnston, December 29, 1948.

Scientific Advisory Committees

- 1995–97 NAS Panel on Atmospheric Effects of Aviation.
- 1988–94 NASA, Science Advisor to High Speed Civil Transport Studies.
- 1989–92 NAS Committee on Atmospheric Chemistry.
- 1978–82 Federal Aviation Administration's High Altitude Pollution Program.
- 1972–75 NAS, Committee on Ozone and Stratospheric Aviation.
- 1971–75 NAS Committee on Motor Vehicle Emissions.
- 1969–73 California Statewide Air Pollution Research Center.
- 1965–67 NAS Panel to National Bureau of Standards.
- 1963–67 Panel Member of President's Science Advisory Board on Atmospheric Sciences.

Figure 8.6. 50 years later. Mr. and Mrs. Harold Johnston, December 1998.

Children:	Grandchildren:
Shirley, 1951	Bryce
Linda, 1953	Sara, Laura, Joseph, Aaron
David, 1956	Miranda
Barbara, 1969	Liberty

Additional autobiographical articles

Harold Johnston, Autobiographical Sketch, *Journal of Physical Chemistry* **A105,** 1388–1390 (2001).

H. S. Johnston, Atmospheric Ozone, *Annual Review of Physical Chemistry* **43**, 1–35 (1992).

INDEX